Creepy Crawlies

and the Scientific Method

Second Edition

For Ken, Sadie, Alan, Matt, and Robin (my family), and for my friend Beth Henry, all of whom have inspired and taught me so much about nature.

For the incredible teachers who've taken part in this project, especially Susan King, Goldie Stribling, and Jennifer Joyner. You've shown me what teaching is all about.

For all students who love living things, especially those who've shared my experiments, and posed for my photos—thank you for caring, thank you for helping me! You are the future!

Text and photography copyright © 2015 by Sally Kneidel

Original copyright © 1993 by Sally Kneidel

Library of Congress Cataloging-in-Publication Data

Kneidel, Sally Stenhouse.
  Creepy crawlies and the scientific method : more than 100 hands-on science experiments for children. -- Second edition.
    pages cm
  Includes bibliographical references and index.
  ISBN 978-1-938486-32-6
  1. Animals--Experiments--Juvenile literature. 2. Science--Methodology--Juvenile literature. I. Title.
  QL52.6.K58 2015
  591.078--dc23
              2014047287

Printed in the United States of America
0 9 8 7 6 5 4 3 2 1

Fulcrum Publishing
4690 Table Mountain Dr., Ste. 100
Golden, CO 80403
800-992-2908 • 303-277-1623
www.fulcrumbooks.com

Creepy Crawlies

and the Scientific Method

More Than 100 Hands-On
Science Experiments for Children

Second Edition

Sally Kneidel, PhD

FULCRUM

Contents

Chapter 1
The Scientific Method

Chapter 2
Attracting and Maintaining Critters

Beloved Butterflies and Citizen Science

Chapter 3
Beautiful and Big: Black Swallowtail Butterflies

Chapter 4
Mystical, Magical Monarchs Need Our Help

Bugs That Squeak and Hiss

Creatures Found In and Under Logs

Terrestrial Predators

Insect Reproduction

Preface to the First Edition

Tiffany is a petite and pretty little girl in my daughter's first-grade class. She has a shy smile, big brown eyes usually downcast, and tiny little braids with pink bows on the ends. When I met Tiffany she had a severe case of the anticrawlies, meaning she had an extreme aversion to any animal that was smaller than a kitten. Especially the nonfurry ones. This is unusual in children who are that young, but nonetheless Tiffany was so affected. The first time I showed her a critter, a salamander I think it was, she recoiled in horror and begged me not to make her look at it. But the last time I saw Tiffany she was running through the woods, flipping over logs and picking up snails and worms. Ten minutes after that she was showing them proudly to her class like some sort of prize and explaining what they eat. You might say Tiffany underwent a transformation. Although she was a hard-core resister in the beginning, Tiffany was one of the children who helped me develop and put into practice the ideas for nature study that I've presented in this book. Most of the children took to the little animals right away. Tiffany took awhile to warm up, but her enthusiasm was unsurpassed in the end. I was excited and inspired by Tiffany's progress and by the eagerness of the other children as well. The ideas for children and students that I've presented in this book are from the perspective of (1) a mother who keeps an insect zoo in her house and serves as a volunteer roving science instructor at her children's elementary school and (2) a former teacher and scientist (PhD) who spent eight years in graduate school learning field and lab ecology. I drifted into ecology for children gradually. I started off just keeping things in my home for my own children to look at. Their friends' mothers often asked me to show their children the mantises, the tadpoles, the roly-polies, the fruit flies, the slime molds, and the worms.

When my daughter started school I found that most of the teachers were open to and interested in bringing "hands-on" science into their classrooms, but were uncertain as to how to go about it. Her first-grade teacher, Susan King, was downright eager to explore whatever I had to offer, and her avid interest fed my own. I began taking things in once every week or two: a lizard, a puffball, a bird's nest. The children were fascinated. Slowly the teacher and I began to go from simply "show-and-tell" to activities and then to actual experiments. First, jack-o'-lantern mushrooms, then slime molds, then roly-polies, then praying mantises. I was surprised that the children so easily understood the concept of the scientific method. What do I mean by "the scientific method"? I mean learning and practicing a five- or six-step procedure that is the underlying format or structure for all scientific experiments in every field from sociology to atomic physics. The scientific method is science. The children became adept at all the steps—coming up with questions, turning a question into a prediction (a hypothesis), and thinking of ways to test the hypothesis. They jumped out of their seats to check the results of their experiments and yelped with excitement when their predictions were right.

I think it's important that we start science early, before children decide that it's too difficult or too dull. They need to learn while they're very young that science is in part a process that's fun, not just facts to be memorized. Science is in part finding things about the natural world that interest you and learning how to answer your own questions about them. How could that be dull?

Of the three basic sciences (physics, chemistry, and biology), I think biology is the one children have the most natural attraction for. Think about the zoo and farm animals that even toddlers love. Children's storybooks are absolutely full of animals, and it's not a very big leap from zoo animals to the smaller versions that can be found under a log. The main difference is in the attitudes children pick up from their elders. I've learned from working with first graders as well as college students that most younger children have little aversion to bugs and worms and such. They almost all ask to hold any creature, whereas college students and adults almost never do. If we can preserve their natural innocence and encourage their interest, then studying critters can provide a real jumping-off point for the study of scientific methodology. Children can learn scientific methodology with living things just as well as with physical or mechanical systems. It makes sense to take advantage of their natural fascination with animals—any animals.

What's great about children's natural interest in animals and about animals' adaptability to scientific study is this: There is a vast array of animals in your own yard, probably in your own house. The last time I took a bug, a dragonfly larva, into my daughter's first-grade class, her classmate Emily asked, "Where do you get all these things?"

"In my yard," I said, "and I don't have a special yard—it's just a regular old yard."

"It is?" Emily exclaimed, astonished.

All of the animals I talk about in this book are things that I found in my own yard, and I live on a regular city lot with a sometimes-flowing ditch behind it. This is not to say that I can walk out my front door and find a praying mantis or a caterpillar or a swallowtail butterfly or a slime mold anytime I want to. Rather, it's a matter of looking for things every time I go outside and trying to find a way to make use of what I find. Most times I don't find anything. But sometimes I do. You may find totally different things from those I find. But if you can catch it, you can use the principles in this book to adapt it to an experiment. I write in the second chapter about how to provide the basic necessities that will keep most critters alive and well for several days, long enough to do an experiment. I also give some tips on how to provide habitats outdoors for critters in order to improve the likelihood of finding certain types of animals.

I learned to look while I was in graduate school studying for my doctorate in ecology, but not through anything that happened in the classroom. Most of the graduate students were single and we spent all our time together, problem solving. Each of us had a different project with a different animal in a different habitat. Michele studied fish, Ellen and Tim worked with tiny aquatic crustaceans, Phil studied roly-polies, Rich studied beetles, Kiisa worked with salamanders, Judy with aphids, Alan with wasps, and so on.

For Michele to work with mosquito fish, she had to know about all the predators of mosquito fish as well as all the smaller animals that mosquito fish eat. She had

to understand the interconnection among all the animals in that habitat. If she or someone else needed help setting up pens for her experiments in the field, or help identifying some of the organisms she'd caught, whoever helped her found out all about her ecosystem as well. We learned an awful lot from each other this way.

The upshot of all this was that I learned that there are fascinating little systems of animals everywhere. Judy studied two types of aphids—one red and one green—that live and feed on goldenrod plants. She found that they are able to coexist, or share their resource, only because their predators (ladybug larvae) keep the aphids' numbers so low that the aphids never are in competition for food. Otherwise one species would outcompete and eliminate the other. Ron found that reducing the density of fish in a penned area of a marsh altered the relationship of all their prey species to one another. The dominant prey species were replaced by others. And so on.

I've discovered that a lot of what I learned incidentally in graduate school is of real interest to little kids. It's been too much fun to keep to myself, and it's valuable to the children. Mrs. King says she thinks that if the children remember anything about first grade, it'll be the hands-on things we did with crawlies.

If all of our children could learn to regard the tiny and usually unseen residents of our natural areas as creatures worthy of study, maybe we would find in the future more effort devoted to preserving what is left of our undisturbed lands.

This book was written in honor of Mrs. King, my daughter's first-grade teacher, who first planted the seed in my mind, and the many other teachers at Cotswold Elementary who have invited me into their classrooms and provided an audience for me. It's intended for all teachers who have an interest and are looking for the means. I specifically have had elementary teachers in mind, but all of the experiments herein are suitable for secondary schools or college biology labs. Homeschoolers and other parents can use it at home as well. There is no equipment required that would be available only at schools.

Most of the animals in the book can be maintained in the classroom for an indefinite period of time. My approach in classrooms has been to take in one type of creature at a time and spend a couple of weeks on that one, then move on to a different one. I keep many going simultaneously at home, but most teachers find one at a time enough when doing experiments.

Preface
to the Second Edition

I've changed during the years since the first edition of this book was published. I'm still a bug and wildlife fanatic, still love working with students, but I'm more concerned about conservation than I used to be. Our planet is not the same place it was when I wrote the first edition. I spend more time with environmental activists these days, working to slow the pace of global warming.

The eager students who participated in the first edition have changed as well, including my two children, Sadie and Alan. They've grown up! And both are avid naturalists. They still like to poke around in the bug cages on our kitchen table when they come to visit! They also share my growing concerns about our fragile planet and its inhabitants, both human and nonhuman. My last two books from Fulcrum, coauthored with Sadie, were both about environmental issues.

Although the environment we all live in has changed significantly during these years, I take solace in knowing that the behavior of the crickets and worms and other old favorites in this book hasn't. Invertebrates are mostly hardwired in how they manage their lives and how they respond to the variables presented in this book. The nature of the scientific method hasn't changed either, nor has the fascination of students with living things. So the basics of this book remain the same, and you'll find most of the popular tried-and-true creatures and experiments still here.

In this second edition, however, I've added four new chapters about four new species. I considered quite a few candidates for inclusion but settled on the four that students like best, which happen to be the same four I like best. I fell in love with some of these new critters just after the original *Creepy Crawlies and the Scientific Method* went to press. I was sad about missing that boat, but I began using these new insect friends in my ongoing science classes. What a blast I've had with them! They're some of the most entertaining bugs I've met in my lifetime—interesting enough that I've kept many of them for years, like other families keep dogs and cats. I even bring them out at dinner parties sometimes, just because

they always generate interest. I like these new bugs partly because students are drawn to them—but also because they're amazing insects with some pretty crazy behaviors!

Along with these charismatic new creatures, I have some new experiments (also too late for the original *Creepy Crawlies*). They're some of my most beloved experiments because they're dramatic and fun for students. That "fun" part is important, because I've learned over the years that students remember feelings and impressions much more than they remember facts. And especially in today's world, any opportunity to create good feelings about living things is crucial. The sense of wonder that children display when they pick up a critter and marvel at its peculiar quirks creates a "teachable moment."

Suddenly, it's no longer a dismissible lowlife but rather a unique and unexpectedly complex individual species, worthy of having its habitat protected. One of my mantras is "out of awareness comes concern, from concern comes action." All living things on this threatened planet can benefit from increased awareness, concern, and action.

Adding four new chapters meant that some older chapters had to go. That was a difficult choice, but, in the end, my decision was based on conservation. We're losing habitat at an unprecedented rate due to climate change, deforestation, draining of wetlands, and pollution. Habitat loss leads to the decline and extinction of species. Human trafficking in animals and animal parts (legal and illegal) is another major driver of wildlife declines.

The human population has increased by 31 percent since the first edition of this book was published, and population growth is, of course, the main reason for all of our environmental woes. Many more species are threatened with extinction than when the first edition was written, and the rate of extinction will accelerate as we move through this century. I worry about that. I worry about it a lot.

Most of the animals featured in the first edition are invertebrates—animals without backbones. The only exceptions are the tadpoles and toads (amphibians) and the mosquito fish, *Gambusia*.

Sadly (for me), I decided to omit from this second edition the two chapters that included amphibians and/ or fish. I love amphibians, but am leaving them out for their own protection. Amphibians are experiencing mysterious and calamitous population crashes around the globe, often within a single year. Although many or most vertebrate species are declining, these sudden plunges to extinction for various amphibian species are especially dramatic. Reasons include the usual suspects: climate change, habitat alteration, pollution. The thin skin of amphibians makes them particularly vulnerable to pollution, but scientists also suspect that a fungus may be a major contributor in sudden collapses of amphibian populations. Humans may play a role by transporting amphibians across countries and continents for the pet trade, traditional medicines, research, food, roadside zoos,

and so on. Transporting an animal means transporting any fungus it may have as well. So, you can see why I don't want to encourage readers to order amphibians by mail, or to diminish local populations by collecting them.

I considered retaining the "Pond Ecology in the Classroom" while eliminating the tadpoles from that chapter, but many of the experiments involve tadpoles. And many of the other creatures in the chapter are not as easy to come by as they used to be. Many of our wetlands have been drained for development. I do live in a rapidly growing city, and most of the places where I used to find aquatic insects no longer exist.

I think you'll like the four new chapters. I know from experience that these are bugs children will remember and enjoy. They'll take away not only a better understanding of scientific experimentation but also a new awareness, appreciation, and love for the diminutive and intricate creatures that share our planet.

Acknowledgments
for the First Edition

I want to thank most of all my husband, Ken, for contributing to every stage of putting the book together. He helped me decide which creatures to use, helped work out ideas for experiments, and "field-tested" several experiments in his high school biology class. His knowledge of insects was a great help at every turn. I thank him also for his editing and feedback on early drafts.

I am grateful to my children, Sadie and Alan, for helping me collect insects, posing for pictures, gladly sharing their house with a multitude of bugs, proudly showing new bugs to the neighborhood kids, and in other ways being interested. The assistance of both children in the classroom was invaluable. Alan was a preschooler and with me full-time when I did most of the classroom work. He carted bugs to and from classes with me and helped feed and water bugs on our daily classroom rounds. Both children passed out materials, contributed information, and shared their zeal with other students. Thanks for the excellent teamwork.

I thank all the teachers at Cotswold Elementary in Charlotte, North Carolina, who invited me and my creatures into their classrooms, where I found my inspiration. In particular I thank Susan King and Goldie Stribling, Kelly Bowers, Belinda Cannon, Veronica Carter, Faith Dunn, Emily Gulledge, and Margaret Rowe. I am grateful to Elva Cooper, the principal at Cotswold, for her encouragement and support. I thank the other staff and the parents who made me feel welcome.

This book would not have been possible without the students at Cotswold Elementary, who amazed me with their fresh and undaunted interest, their bright eyes and shining smiles, their little hands so ready to receive worms or bugs or toads, and their unending questions and willingness to learn. The steadfastness of their interest surprised me and fueled me every step of the way.

I thank all of the children at Cotswold Elementary who allowed me to photograph them in action, and their parents for giving me permission to use the photographs.

I am grateful to fate for having run me past Sam Watson, whose message to me about writing made a difference. Thank you, Sam, for persisting.

I am indebted to all of the people I knew in the Biology Department and the Ecology Curriculum at UNC at Chapel Hill from 1979 to 1984, who taught me about the natural world and the scientific method. Any errors in the book are my own responsibility.

I want to express my gratitude to my agent Sally McMillan for her efforts in my behalf and for her steady support and interest. I'm grateful too to the staff at Fulcrum Publishing, especially Carmel Huestis and David Nuss, for their guidance, encouragement, and editing.

Thanks to my longtime friend Kathleen Jardine for her continued faith in me and her appreciation of all the little animals and fungi we examine together on our walks in the woods.

And last but not least, I'm grateful to the insects—the tadpoles, millipedes, spiders, and worms—for being there so predictably and for surprising me and amusing me in so many ways.

Acknowledgments for the Second Edition

Again, I want to thank most of all my husband, Ken, for supporting and encouraging me with the second edition. He helped me explore which new creatures to try and helped me work out details of new experiments when I hit a snag. His overall knowledge of insects and science was, as usual, a great help. I also appreciate his tolerating the amount of time I spent around the clock on putting together the second edition.

I thank my friend Beth Henry for inspiring me and encouraging me in so many ways, for sharing her remarkable knowledge of butterflies and native plants, and for turning me toward activism.

Thanks to my daughter, Sadie, for helping me with her always insightful editing, for understanding my preoccupation with this project, and for encouraging me.

I'm grateful to my son, Alan, for his lifelong expertise and mastery of butterflies, as well as birds. His enthusiasm is contagious!

I am greatly indebted to Jennifer Joyner, an extraordinary teacher and monarch expert. She really showed me what can be done with monarchs in the classroom and brought the options to life.

I want to express my appreciation to all the children who tested my experiments, loved my bugs, and agreed to be photographed. You were all essential, and thank you! Especially Lily and Nia!

I am grateful once again to my agent Sally McMillan for all her good assistance and advice over the years.

Thanks to Richard and Pam for keeping me afloat in so many ways. I love you!

Introduction

There are many books that can help children study nature, but this one is unique in a couple of ways. First, it contains 114 experiments, mostly behavioral, with animals that are commonly found in nature. Each experiment is a five-step procedure: **question, hypothesis, methods, result,** and **conclusion.** Chapter 1 is devoted to explaining these five steps, which together constitute the scientific method. So this is more than just an activity book; these are real experiments in the academic sense.

You can, however, use this as an activity book. The experiments are the last part of every chapter. The first part of each chapter describes the animals (mostly insects), how to find them, how to maintain them, and things you can do to encourage observation. There's a lot of enjoyment to be gleaned from just keeping the animals for observation and letting the children hold them in most cases. This is especially fun when each child has his or her own. So each chapter provides the option of stopping short of the experiments if you wish.

A second unique feature of this book is that I have tried to make it more than just an instruction manual. I've tried to convey my own appreciation for these miniature living things. Sharing them with children has helped fuel my own delight in them. Each type of creature is peculiar in its own way; each is so different from ourselves, yet similar too. They all need food, water, and shelter. And even the most lowly is capable of making choices about its habitat or food.

I've included students' reactions when I think they would be useful, and I've tried to give readers the benefit of my own experience of getting things to work out right in the classroom. It's not just an instruction manual—I mean for it to be a map into a tiny wonderland. My small students with their open minds and their ability to marvel at the simplest things have helped me to appreciate that perspective.

My experience with elementary students has all been as a roving science teacher. I take the creatures into someone else's classroom, set things up, and turn over the ongoing care to the students and their teacher. Then I come back periodically for results and feedback.

The Format of Chapters 3–18

Following is a list and then a description of the sections in each of the experimental chapters that I carried over from the first edition, Chapters 7–18. The new chapters, Chapters 3–6, may vary slightly from this format. Chapter 4 about monarchs is significantly different from the format shown here because it doesn't involve experiments—although it does include plenty of activities! I include the following list and description because it might be helpful to familiarize yourself with the predominant structure in the book before reading.

Introduction

Materials

Background Information

How to Get and Keep the Organism

Field Hunt

The Organism at School

Getting Ready

Observations and Activities

Experiments

Introduction Explanation

The short introduction to each chapter is a summary and also usually tells you what I consider to be the most remarkable feature of that particular animal.

Materials

None of the materials are very expensive, and in many cases you can make do with things you may already have on hand. It will be more fun, though—and easier—if you order a few supplies. The materials I use most frequently are fruit fly culture vials, petri dishes, plastic peanut butter jars, and insect sleeve cages. These are described in Chapter 2 and can be ordered from a biological supply company or bought at a science hobby shop.

Although many of the animals in this book can be caught easily outside, some can be ordered too. The materials list or the How to Get and Keep the Organism section in most chapters will tell you if this is possible. You'll find a list of biological supply companies in the Appendix; any place in the book where I suggest ordering something refers to these places, unless stated otherwise.

Most of the chapters have a materials list that will tell you which materials are needed for which experiments. Commonly available materials such as tape or scissors are not listed.

Background Information

This section tells you about the life cycle and natural history of the chapter's subject. Certain aspects may be interjected later, just previous to relevant experiments.

How to Get and Keep the Organisms

Here you'll discover how to go about catching your subjects in nature. I've provided very detailed accounts of how to feed and house the creatures. Their daily maintenance is probably the most pleasurable aspect of the whole enterprise to me (other than the look of delight on children's faces). I can spend hours rearranging cages and watching animals eat, hours when I'm supposed to be doing other things. But daily maintenance need not take more than a few minutes. Many don't need attention every day.

Field Hunt

This section of most chapters gives information on how to take a class out searching for the particular creature. This can be a fun and worthwhile activity in and of itself, not necessarily leading to classroom experiences. A damp paper towel in the collecting container will keep captive insects and other invertebrates happy for a few hours. The inexpensive *Insects, Revised and Updated* (Golden Guide) by Zim and Cottam will help identify most insects encountered.

The Organism at School

Here I describe how to set up your class to receive the creatures, and usually I explain how I've introduced students to the animals. Where I think it would be useful, I've given children's reactions and questions. I continue to be amazed at the consistency of their enthusiastic interest. And it's not me—I'm a very quiet and shy person, inept at speaking in front of groups. It's the creatures! Kids love 'em! I took some ant lions to a third-grade class this morning. Lots of kids in the class had worked with ant lions the year before, so I thought they might be indifferent. But it didn't seem to make any difference. They were leaning out of their seats to see, and half the hands in the class were up wagging with comments and questions the whole time I was talking.

The Organism at School section also describes how to encourage observation. With some animals, one option is to let each child have his or her own animal in a small container at his or her desk. With others, like adult praying mantises or water bugs, this is not practical. In some chapters, I suggest activities to help the students notice certain features of the animals.

Next come the experiments. The number of experiments per chapter varies. Some chapters include tables you can photocopy for the students to fill out, or graphs on which you can plot your own counts or measurements. The Appendix includes a blank graph and table you can photocopy if a chapter does not provide one that fits your needs. In some chapters completed graphs are included to provide examples of how a graph might look. Your graphs may look very different from mine. Conditions vary and animals are not always predictable!

The Postscript following Chapter 18 ties together common threads among chapters. At the end of the Postscript are tables and lists of questions that will help students compare results from similar chapters. For example, one table compares responses of under-log creatures, from the first group of chapters. A list of questions compares strategies of different predators, from the second and third sections of the book.

My hope is that reading through the book and trying some experiments will help teachers and students come up with some of their own questions that can be tested. Not just any question can be made into an experiment—I'll elaborate on that in Chapter 1. All it takes is practice to fall into that questioning mode of thinking. Then you're off and running.

The Appendix supplies a list of sources for invertebrates and supplies.

Chapter 1
The Scientific Method

Why Do an Experiment?

Why not just keep animals in terraria in your classroom and observe them? I spend a lot of time just watching animals, feeding them, talking to them, and in general regarding them as pets. It's fun and it's valuable. I don't routinely do experiments with the animals that live in terraria around my house. I just enjoy them.

The classroom is a different situation. Some children will spend a lot of time watching gerbils or fish in a classroom. Others have less interest. But when a child makes a prediction about that fish's response to something, like covering half of the aquarium with a dark towel, that child invests something of him- or herself in the fish's behavior. It may greatly increase the child's interest in the fish. Say the child then observes that this particular type of fish avoids the dark part of the tank. If the child writes down his or her discovery and shares it with others, the child's going to feel pretty proud. Most of the projects I did as an undergraduate biology student consisted of simply observing captive animals' behavior. I learned a lot from doing that and children will too. But children will find it more fun, more interesting, to make predictions and test them—or in the case of the monarch butterflies, to engage with the scientific community in conservation efforts.

Another important advantage of hypothesis testing is that posing questions, predicting answers, and testing the predictions all develop higher-level thinking skills. Experimentation develops the habit of following observations with questions, a trait that adult innovators all possess.

A third and perhaps most important advantage of experimentation over observation is that by conducting experiments, children learn the meaning of the word

I didn't feel like a scientist until I started doing experiments.

science and think of themselves as scientists. I never did any science at all in biology until I got to graduate school. That is, I never did an experiment until then. I read a lot and I learned a lot, but I didn't feel like a scientist until I started doing experiments.

What was the big deal about doing experiments? I learned that instead of memorizing stuff other people had discovered, I was creating knowledge. I was contributing to the accumulated body of knowledge. I understood then that science is not an already existing set of facts but an ongoing process of discovery, and I had become a part of this process. By teaching children to do experiments, we give them that feeling too—that science is an ongoing process and they are part of that process. It's an exciting feeling. It ceases to be intimidating and becomes a source of pride.

Okay. But how does a nonscientist develop scientific thinking in a classroom? Science is not necessarily complicated thinking. It's just a method of addressing questions. In my ecology program all of us began by simply picking an area of interest. We then spent time in the field poking around in a pond, or a tree hole or a stream—wherever we were likely to find the creatures that interested us. If we could we'd take several back to the lab to keep in captivity and observe continually. That's how questions occurred to us. If there is no period of observation, then no questions will occur. This can be a fundamental stumbling block if children try to come up with experiments with no prior observation period.

All of the experiments going on in my department at graduate school were simple. (What gives length to the dissertation is the survey of related research and perhaps a series of simple experiments.) Most experiments were based around a simple yes or no question. For example:

Do spotted salamanders survive longer with predators if they have weeds to hide in? (Yes or no.) This was one of my experiments.

This has what seems like an obvious answer. But in science, even an outcome you can guess must be demonstrated experimentally if it is to be accepted by one's colleagues.

Many science projects and many experiments have obvious outcomes, but they are still worth doing. Once you have a simple yes or no question, it's easy to make a hypothesis. Simply state what you think the answer is. Or have the students state what they think the answer is. For example: I think mosquito larvae will survive longer if there are weeds in the water.

Here is an important point: The hypothesis must be testable. The above hypothesis can be tested by putting some mosquito larvae and their predators in an aquarium with weeds and putting others in an aquarium without weeds. Then after a certain length of time, I remove the weeds and count how many are left in each aquarium.

Here is an example of a hypothesis, based on a yes or no question, that is not testable.

Question:

Do slime molds like warm dishes better than cold dishes?

Hypothesis:

Slime molds like warm dishes better than cold dishes.

"Like" is not something we can count or measure. A testable hypothesis must predict an outcome that can be counted, measured, or concretely observed in some way. With slight modification, the above can be made into a testable hypothesis: Slime molds survive longer (or grow bigger) in a warm environment than in a cold environment.

These are the steps of the scientific method, as applied to teaching children.

1. Observation.

Always begin with a period of observation (unless you already have a question). One observation might be, "Roly-polies are found in dark places, like under bricks."

2. Question.

Get the children to ask questions and focus on one question. For example, "Do roly-polies like to be in the dark?"

3. Hypothesis.

Get the children to predict what they think the answer to the question will be. For example, "Roly-polies are afraid of the dark." Make sure the hypothesis is testable, which the above is not. Many predictions that children make spontaneously will not be testable, since testing a hypothesis should produce a measurement or count or an objective observation. A feeling can't be counted or measured or indisputably observed, but most behavior can. The above prediction can be changed to, "Roly-polies will choose light over dark." Most predictions can be modified in some way to be made testable.

Some teachers prefer that students make predictions as "if" statements. For example, "If I offer roly-polies a choice between light and dark, roly-polies will choose dark."

4. Methods.

Carry out the experiment. Give the roly-poly a choice. You may want to replicate your experiment (see Replicates, p. 19).

5. Result.

State exactly how the roly-poly reacted and which side it chose. For example, "The roly-poly wandered around the light side for five minutes, trying to crawl up the side of the container. Then it entered the dark side and stayed there until the predetermined time to end the experiment."

6. Conclusion.

Here you state whether your prediction was confirmed or not and try to explain your results.

The Null Hypothesis

Scientists try to make sure their results are not biased by their expectations. One way they do this is to try to disprove the hypothesis, rather than try to prove it. Another way of going about this is to state a "null hypothesis." Instead of stating what effect you think your experimental variable will have on the animal, you state as your hypothesis that the variable will have no effect. For example, a null hypothesis would be, "We think cold will have no effect on the growth rate of mealworms," when actually you expect that cold will make them grow more slowly.

In my experience, young children are more successful at coming up with a hypothesis when they're allowed to say what they think will happen. Trying to convert their natural prediction to a null hypothesis can be confusing; I leave this up to the individual teacher.

Replicates

If you had one roly-poly and recorded its choice of light over dark only once, you would not have much confidence that roly-polies always prefer darkness. Consider this example. Suppose you drink coffee most mornings, but occasionally drink tea. A coffee company representative calls you Tuesday morning, doing a survey, and asks if you are drinking coffee or tea. You say tea. If she concluded you always drink tea she'd be wrong. She'd be even more wrong to use this to conclude that everyone always drinks tea.

It is easier to generalize from one roly-poly to the next than it is from one person to the next. Roly-polies' preferences are genetically programmed, so if one of them definitely prefers dark, then probably all of the same species do. But it could be that roly-polies don't really have any preference between light and dark. If not, then giving one roly-poly a choice would be like flipping a coin. You could flip it once. Or you could even flip it three times and get three heads in a row. It's not very likely but it's certainly possible to get three heads in a row by chance. And if you gave your roly-poly a choice between light and dark three times in a row, it could choose dark three times in a row just by chance.

How many times would you have to offer your roly-poly a choice to be sure that your results were not due to chance alone? If I flipped a coin ten times and got heads nine out of ten times, I would wonder if the coin was weighted on one side. If I flipped the coin fifteen times and got heads thirteen out of fifteen times, I would be sure the coin was weighted. I would be sure the results were not due to chance alone. And so, to be sure that my results are really answering the question, I might want to offer the roly-poly a choice ten times.

Each repetition of an experiment is called a replicate, or a trial. The more replicates you have, the more confidence you'll have that your results reflect the truth. With children, though, scientific "truth" may not be our highest priority. Sometimes when I get too rigorous with children I lose my audience. We want the children to understand the process and to enjoy themselves. If more replicates become tedious, then keep it simple. The concept of how to go about answering a question may be more important than the veracity of the results in some situations. You can always discuss how additional replicates could affect your results and conclusions if you employed them.

It's important too if you're doing a series of experiments to decide before you start how many trials or replicates you're going to do. This is so your results will not be influenced by your expectations. For example, suppose I think my coin is weighted on one side and I want to prove it. I could say, "I'll just keep flipping until I have more heads than tails." Well, during the course of my flipping, the ratio of heads to tails will seldom be exactly 1:1. It fluctuates. After four flips I may have three heads and one tail. But after twelve flips I may have five heads and seven tails. The balance fluctuates. If I choose when to stop after I've already started, I can choose a stopping point that supports my expectations. I'll stop after four flips if I want to prove that the heads side is weighted.

Variability in Results

If you conduct an experiment two times with animals of the same species, same age, and same gender under precisely the same conditions, you should get precisely the same results. In an elementary school setting, however, it's difficult to control everything precisely. So your experiences may vary somewhat from those I discuss in the book. You may use different species whose behavior varies somewhat. Not all roly-poly species react to light the same way. There are several species of crickets called field crickets. All are territorial, but their courtship behaviors probably vary slightly. There are many species of mantises in the United States too.

Small changes in environmental conditions can make a difference. In a particular experiment, your sand may be wetter or drier than mine. Your maze may be made of different materials.

The behavior of individual animals can vary too. In Chapter 5 I describe how the level of aggression in a male cricket depends on its recent experiences with territory and other crickets.

Variability can be fruit for discussion and further experimentation. If your results vary from what my comments lead you to expect, help the students speculate about why. It may lead to another experiment: "Let's try it again with drier sand."

Controls

A control is simply a part of the experiment that gives you something to compare your result to. An experiment for which an animal is offered two options, like the one described earlier with roly-polies, already has a control. If I predict that roly-polies will choose light, then offering dark as well is my control. I can compare their reaction to light with their reaction to dark. If I did not have darkness as an option, I wouldn't know anything about their preference for or aversion to light.

But not all experiments have a built-in control like that one. Suppose you want to know whether mosquito larvae are able to survive longer with predators if there are weeds in the aquarium. This requires a control. Say you put ten mosquito larvae in an aquarium with pondweed. You added predatory dragonfly larvae. Suppose you had six surviving mosquito larvae after twenty-four hours. This wouldn't tell you a thing about whether the weeds helped the mosquito larvae escape predation. Some have escaped so far, but why? You don't know for sure. You need something to compare your result to—an aquarium without weeds. If more mosquito larvae survived in the aquarium with weeds than in the one without weeds, then you'd have some information about the effect of weeds. So the aquarium without weeds is your control. It allows you to attribute your result to the presence of weeds.

The Experimental Variable

It is important that there be only one difference between the control and the experimental setup. In the last example the only difference between the two aquaria is that one has weeds and one has no weeds. In other words, weeds are your "experimental variable" because you are testing a prediction about the effect of weeds. All other factors must be equal between the two aquaria, like size and type of predator, volume of water, illumination, disturbances, and so on. Any factor that could affect the outcome must be the same between the two aquaria—except the experimental variable.

How Is an Experiment Different from an Activity?

An activity is simply watching something, or perhaps interacting with it in some way so as to cause a reaction. For example, feeding a live cricket to a praying mantis is an activity. Many people call activities experiments. But an experiment is an activity that is designed to answer a question and has a control to rule out other interpretations of the result. An experiment is more valuable to a child's learning because it encourages more thinking. An activity is really part of the observation period and can lead to a question and an experiment. Say we've got a praying mantis in captivity. We've read that it eats insects, so children bring in insects—some are alive but most are dead. The children notice that some things are eaten and some aren't. None of the dead insects are eaten. Some of the live ones, but not all, are eaten. Some insects that sit very still in the cage, like moths, are not eaten. At this point some questions can be asked that lead to experiments. Why does the mantis not eat the dead things? Why does it eat a live cricket but not a live moth? What is the difference between the cricket and the moth? Does the mantis eat crickets because they move more? Coming up with questions like these requires critical and analytic thinking. Children become more proficient at this with experience.

You've got a question now that will invite predictions or hypotheses. For example, a mantis will eat a moving insect before eating a nonmoving one. The method of testing this prediction can be to offer other pairs of moving and nonmoving prey, like a housefly (moving) and a grasshopper (mostly nonmoving in captivity). Or a regular cricket and a cricket that has been chilled for an hour or two in the refrigerator to paralyze it temporarily. Or jiggle a dead cricket with tweezers in front of the mantis.

This can lead to further questions and hypotheses. Will a mantis eat anything that moves? A jiggled piece of ground beef? A jiggled piece of apple? A jiggled piece of paper? How much movement is required, continuous or only once?

Why do an experiment? Why not just offer the mantis a variety of items, record what it eats, look for patterns and generalizations, and leave it at that? The value of making predictions is that the students invest themselves in the outcome. This greatly increases interest. And by doing an experiment with a control, children learn the process that is the foundation of all our knowledge in science. The children become a part of that process by contributing to that body of knowledge. They become scientists.

Chapter 2
Attracting and Maintaining Critters

Some of my best experiences in working with students have been field hunts for critters on the school grounds. The children love it. A field hunt can whet their appetites for keeping animals in the classroom and also helps them relate their school experience to home, where they can explore at leisure. Most chapters have a Field Hunt section that describes how to find in nature the animals focused on in that chapter. What I've described here is not so much how to attract or find anything in particular, but just how to provide fun places to look.

I've described here the cages that I find most useful. Terraria are too big for some things. An interesting cage can motivate even a squeamish child to find something to put in it. Those I suggest for general bug catching are the same ones you'll need for many of the experiments in this book.

Fig 2.1 A land planarian. Most encountered in nature are less than 8 inches (20.3 cm) long. They are brown with black stripes running the length of the body. The shovel-shaped head distinguishes them from other worms.

The Under-Log and Under-Stone Crowd

A sheet of plywood or dark plastic left lying (and anchored) on damp ground, preferably with a little dead vegetation underneath, will almost certainly attract something. Among the possibilities are crickets, earthworms, roly-polies, millipedes, centipedes, snails, slugs, and small snakes. Spiders are another possibility, so students should be cautioned about not sticking hands in places they can't see into.

Large stones in damp places yield similar results. You may get colonies of ants under large rocks that have been left undisturbed for several weeks or months. If you're lucky you may find a land planarian—a flat, striped, sticky worm with a triangular head, about 8 inches (20.3 cm) or less in length (see Figure 2.1). My kids find them in our suburban yard about four times a year, usually under bricks or stones. I saw several out in the open once at night on a slate walk when I was a child. Their heads and necks are held up in the air when they're on the move, which is very peculiar looking. They ensnare their prey with slime and leave a slime trail behind as well.

My biggest success in drawing large numbers of creatures has been with a pile of wet and very rotten (stinking) grass clippings dumped from a plastic bag onto a gravel driveway. The bag had been closed and in the sun for several days. After I dumped it, I left it untouched for about three weeks. Then when I looked under it, I found fifty to one hundred earthworms, crickets, millipedes, and beetles. I think the spaces in the gravel enabled them to get under the clippings more easily. But after I'd checked under the pile on three or four successive days, the worms and crickets were all gone. Those not captured had fled.

Frequency of disturbance is a major factor in keeping any under-board type creatures available. They'll leave

Fig 2.2 Alan looking under a log.

permanently after being disturbed several times in rapid succession. We used to have a tiny worm snake (the most charming of all the snakes in my opinion) that lived under a concrete slab behind our house. It was there every time we looked as long as I could keep the peeks spaced a month or so apart. But as soon as my kids got big enough to lift the slab, the snake left.

Rotten logs provide a little more variety than plastic sheets and boards left on the ground. You can look under the log (see Figure 2.2) or inside the log, although a log can be ripped apart only once. You'll see wood roaches and a variety of beetles, as well as all of the same things you see under a board or plastic sheet. Under the bark of rotting logs, termites are common, as are ants. Both travel in tunnels that permeate the log, often the same log, but their tunnels never intersect. I love to watch termites. They look something like white pudgy ants. The ones you see in logs aboveground are all workers and soldiers on foraging expeditions for the queen and her young, who usually live underground. You can distinguish the soldiers and workers easily by the much larger jaws of the soldiers. The soldiers' job is to protect the workers while they collect food, and they'll try to bite anything that gets in the way. A soldier bit me once as I collected some termites to feed some captive animal, and I had to laugh at its determination. The bite was barely detectable, but the little bugger would not let go! Termites are blind so it had no way of knowing it was

biting a 130-pound giant. The bite of a soldier *ant*, on the other hand, can be painful—more like a sting for some species of ants.

A vial and a small paintbrush are useful for collecting termites and ants. Both can be kept in captivity temporarily but only the queens can reproduce. You're unlikely to find a queen—anyway, I never have.

Termites eat rotting wood. Different species of ants eat different things. You can order an ant farm kit from a biological supply company (see Appendix) or buy one at a science hobby shop. Termites are vulnerable to dehydration; a termite in a container must have a damp paper towel to survive for more than a few minutes.

Keeping Invertebrates in Jars Temporarily

In general, any insect or other invertebrate captive can be kept alive in a jar for few hours if you keep a wet (not dripping) paper towel in the jar with it. Be careful that there's no water pooling in the bottom of the jar. The paper towel should be crumpled but not balled up, so that the creature can crawl into its nooks and crannies. Most creatures that children collect in jars die of dehydration. The wet paper towel will keep them from losing moisture through their skins or exoskeletons.

Keeping Invertebrates for More Than a Few Hours

It's not a good idea to keep any wild vertebrate animal in captivity for even a few minutes. For conservation and humane reasons, students should get the message that vertebrates need to be left where they were found. A vertebrate is anything with a backbone: fish, amphibians (toads, frogs, salamanders), reptiles (snakes, lizards, turtles), birds, or mammals.

If you want to keep an invertebrate (insect, worm , spider, roly-poly, snail, etc.) for a couple of days, try to duplicate its natural habitat in an enclosure. If it was on a stem, put a stem of the same plant in the enclosure. If it was under a rock, put damp soil and a hiding place in the enclosure. If it was in a log, get a piece of rotting log for it. Damp soil and damp leaves or moss instead of a paper towel can help provide moisture in terraria, and at the same time, keep it looking natural. Many invertebrates are accustomed to drinking dewdrops, so spray the terrarium with a plant sprayer daily.

If you don't know for sure what the invertebrate eats, return it to where you found it after a day or so.

Traps and Cages for Invertebrates

Funnel Traps

A funnel trap is for flying insects. It consists of a closed container with one opening into which a funnel is fitted, so that the small end of the funnel extends into the interior of the container (see Figure 2.3). There must be some sort of bait in the container. Bait can be fruit, meat, or other food, or at night a flashlight. Flying insects that are attracted to light enter the container through the funnel and then can't get out. With food as bait you'll get mostly flies and flying beetles. A small piece of chicken liver or raw fish will attract a lot of big flies you can use to feed a big mantis. With a light as bait, you'll get moths as well as flying beetles and other insects seen on porches at night.

Having an assortment of interesting cages can contribute to children's enjoyment of little invertebrates. Petri dishes and fruit fly culture vials are useful for almost anything in the less-than-1-inch (2.5 cm) range (see Figures 2.3 and 2.4). Both can be ordered from a biological supply company (see Appendix) or found at science hobby shops or on the Internet. Petri dishes come in various sizes, either glass or plastic. I prefer the $3\frac{1}{2}$-inch (9 cm) size in plastic. Fruit fly culture vials are about 4 inches

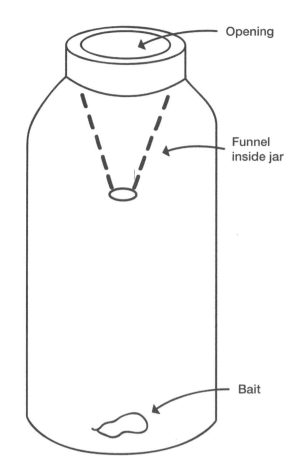

Fig 2.5 A fruit fly culture vial, about 4 inches tall.

(10 cm) tall, plastic, and they come with foam-rubber stoppers. Whenever I refer to vials in this book, I mean fruit fly culture vials. Children will enjoy using these two all-purpose containers to house anything and everything—a small slug from the bus, a roly-poly from under a brick, a tiny spider from the wall. Both petri dishes and fruit fly vials are clear for good visibility and shatterproof (petri dishes will crack with pressure, though), and they can be easily slipped into a pocket. A rubber band will keep the petri dish closed. Some children are more interested in the cage than the bug! But that's okay—scientists are sometimes the same way! Enjoying the paraphernalia is part of enjoying any job or hobby.

If you can provide one or two petri dishes and one or two empty vials and stoppers for each child, you'll be well prepared for many of the experiments in this book.

For invertebrates too big for petri dishes or vials I use clear plastic peanut butter jars or rice jars, the 1-quart or 1-liter size. A cloth lid secured with a rubber band over the top permits airflow.

Students can make a bug house by cutting a window in a milk carton and covering the window with window-screen mesh or plastic wrap. They can clip the top

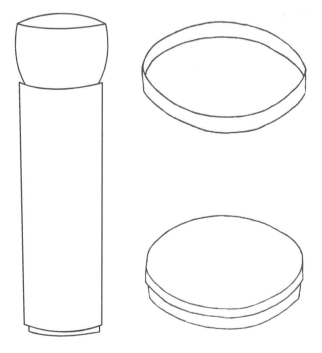

Fig 2.3 A funnel trap. This one is a canning jar with a two-piece metal lid.

Fig 2.4 A petri dish without a lid and with a lid. Clear plastic, about 3.5 inches (9 cm) in diameter.

Fig 2.6 An insect sleeve cage, with a mesh top that has an elastic border. A hand can go into the cage through the cloth sleeve, which stays tied shut when not in use. The cage is about 1 foot (30.4 cm) in diameter.

closed with a clothespin or paper clip, but a lot of small bugs can still escape from these milk carton cages.

A handy cage to have for flying or jumping insects is the insect sleeve cage, which is about the size and shape of a hatbox (see Figure 2.6). This can be ordered from a biological supply company or by searching the Internet (see Appendix). I have three, and all three are always in use. You can stick your hand in through the cloth sleeve to feed and water the animals, without allowing any escapes. You can make one by cutting a circular hole in the side of a plastic bucket and stapling a sleeve around the opening of the hole. If you cover the top with plastic wrap, the animals can still breathe through the sleeve. Those you buy come with a removable mesh top, which I sometimes replace with plastic wrap. Insect sleeve cages have handles and are lightweight for easy transporting.

A terrarium is hard to move around but is preferable to the insect sleeve cage in terms of visibility. (Animals in the insect sleeve cage must be viewed from above.) A terrarium is probably the better choice for animals that won't escape easily and don't need to be moved often.

Beloved Butterflies
and
Citizen Science

Chapter 3
Beautiful and Big:
Black Swallowtail Butterflies

Figure 3.1 Lily releases a newly emerged eastern black swallowtail outdoors. Students love raising caterpillars and then setting them free as adult butterflies.

Bugs with Charm Galore

Butterflies are beloved by children. That's evident from the growing popularity of "butterfly pavilions" where children can walk and schmooze with butterflies that land on their heads or shoulders. I've never seen a child recoil from a butterfly—even a child whose parents hate insects. I have seen children vie with one another to hold a butterfly or a caterpillar, or to release a newly emerged butterfly outdoors. I've seen children squeal with glee when a butterfly takes off on its own. Butterflies are beautiful and delicate. They've been in so many tales of fairies and fantasy that we do, culturally, imbue them with vaguely magical qualities, I think. I have two friends who believe butterflies they encounter outdoors are signs from a higher being. I don't go that far—I believe they're just insects! But they are lovely, and mysterious.

Parents tend to like butterflies too. Butterflies are, I imagine, the most popular insects in the world. Adult butterfly fans sometimes keep "life lists" of species they've seen in nature, with date and location. The area where I live has a listserv for butterfly watchers, where people can post details about rare sightings so that other fans can fix their binoculars on an uncommon species. Their enthusiasm is comparable to the ardor of bird-watchers. In fact, some binoculars are designed and marketed just for butterfly watching.

Why I Chose Eastern Black Swallowtails

Most butterflies are charismatic, but choosing which ones to include in this second edition was easy. I chose eastern black swallowtails because they're dramatically big, common, easy to attract, and easy to raise. I don't personally know of another butterfly that's both so convenient to study and beautiful. They lay their eggs in our garden every year, whether we want them to or not!

Fig. 3.2 Black swallowtails are big and beautiful.

Natural History

Their Distribution

Eastern black swallowtails (EBSs) are found over the eastern two-thirds of the United States. The very similar western black swallowtails (WSBs) are common throughout the West. I live in North Carolina, so the black swallowtails I've raised and shared with students are EBSs. Much of what I write, though—other than the host plants for the larvae—is also applicable to the WBS.

Taxonomy and Name

The swallowtail family Papilionidae has at least 550 species, distributed over every continent except Antarctica. The world's largest butterfly, Queen Alexandra's birdwing, is included in this family. Most of the largest swallowtails live in the Asia-Australia region of the globe.

Why the name swallowtail? Many species in this butterfly family have a long, fingerlike projection off the back of each hind wing. When the wings are open, the projection gives them a forked appearance—which looks somewhat like the forked tail of birds called swallows. Not all papilionid butterflies have the long projections, but black swallowtails do.

Size and Coloration

The wingspan of the eastern black swallowtail ranges from 3.1 to 4.3 inches (8–11 cm), smaller than some of the other swallowtail species in the eastern United States. But still big! The wings of black swallowtails are mostly black. When the wings are open on an EBS, the portion nearest the head is solid black—whether male or female. The adult female has a row of yellow spots along the entire trailing edge of the open wings and a band of iridescent blue spots near the center of the trailing edge. A male EBS also has a row of small yellow spots along the trailing edge of the open wings, similar to a female's, but in addition, he has a row of bigger yellow spots about $1/2$ inch away from the trailing edge. The male has either no blue or less blue than the female on the top of the open wings.

The wings of both sexes when closed are still black on the portion toward the head, but away from the head they have various colored spots too intricate to describe here. Do an image search on the Internet! Or try to identify it with open wings!

Look-Alikes

The eastern and western black swallowtails are not the only swallowtails in the United States that have black wings. Female tiger swallowtails, which we think of as primarily yellow, can be black and often are. But they are bigger than black swallowtails and lay their eggs on different host plants. Likewise, in my area, pipevine swallowtails are black, but again they have different host plants. You can find good pictures of these look-alikes on the Internet.

At any rate, you don't need to be on the lookout for an adult to do the activities and experiments in this book, so identifying adults is not essential—just fun. We start with the eggs or the caterpillars.

Materials You'll Need

Cages and food for eastern black swallowtails are described in the following section. In the latter part of this chapter, each experiment will list any extra materials you may need for that particular experiment.

How to Get and Keep Eastern Black Swallowtails

To find EBS caterpillars, you'll need to have access to one of the plants these particular caterpillars eat (called host plants). Plants that are easy to grow for the caterpillars include parsley, fennel, dill, and carrot greens—all of which are in the carrot family (Apiaceae). EBS caterpillars will also eat rue, in the family Rutaceae. Parsley is the most commonly grown of these in home gardens. You might be able to access parsley plants in a friend's garden, or of course you can plant parsley yourself at home or school—or maybe some students will have parsley in their own home gardens and can bring the caterpillars to school.

Growing Parsley

I prefer parsley for attracting EBSs because it's so easy to find in gardening stores, it's easy to grow, and it has multiple uses in cooking as well. Fresh dill, fennel, and rue don't provide as much edible foliage for the caterpillars and are less available or unavailable at groceries if you run out in the garden. (Grocery purchases must be organic because pesticides can kill caterpillars.)

To get started, you can buy parsley bedding plants (already sprouted) or grow it by seed. I grow it from bedding plants because this method produces big plants much faster. Seeds will give you far more plants for less money, but may take three weeks or more just to germinate.

If you don't have a garden, one option is to plant the young plants or seeds in pots. Place the pots outside in full or partial sun. You have to water outdoor pots more frequently than you would a garden; pots dry out quickly in the sun, especially small pots. Of course, don't spray the parsley with any kind of pesticides. In my city, we can get parsley plants at hardware or gardening stores. Ask employees which of their varieties grows fastest and produces the most foliage. (In my opinion, curly parsley has the most foliage per plant.)

If you don't want parsley, fennel would be my second choice. Some EBSs can be a little finicky about dill. Carrots take much longer to make big greens than parsley since you have to start carrots from seed. Rue produces less foliage than parsley, and if you run out of rue, the caterpillars may be reluctant to switch to a carrot-family plant. That's happened to me, and I had to drive across town repeatedly to get rue from a friend's garden. I'll never use rue again for my primary host plant.

When to Plant the Host Plant

The ideal time to plant parsley is spring. If you want caterpillars in September, then the plants will have plenty of time to grow to an optimum size. Plant seeds outside by sprinkling them over the top of the ground. Keep them moist. After they've sprouted and rooted, thin them. You can plant in midsummer, but intense summer heat is challenging for seedlings and young plants.

If you want parsley and caterpillars in May, you can hope that your previous summer's plants survive the winter (possible in some areas). They are more likely to if you pick off all the flower buds you see in late summer. (The plants die after going to seed.) If you need to plant for May, sow the seeds on potting soil indoors about six weeks before your local climate permits transfer outdoors. The seeds are very small; no need to cover them with soil. Keep them moist by misting.

Parsley is a cold-tolerant plant and can be planted or moved outside at the same time as vegetables than can survive a late frost (beets, Brussels sprouts, carrots, collards, kale, parsley, spinach, etc.). This date varies by location, so ask at a local gardening store. It's usually about a month before the customary date for planting tomatoes, corn, and other vegetables that are killed by frost.

You can read more about growing parsley on herb-gardening websites, just by searching for "growing parsley."

Alternative Host Plants for Experiments

If you want to do one of the experiments later in this chapter that involves moving caterpillars to alternative host plants, you'll need those alternative plants. One option, when the time comes, is to buy organic carrots with the green tops attached, fresh fennel, and fresh dill. You can probably find at least some of these at grocery stores, especially stores that specialize in organic and whole foods. You may find some at gardening stores as small bedding plants. If you find fresh fennel or dill at a grocery, put the cut stems in water and they'll last at least a few days. Carrots with greens attached should last a few days in the refrigerator.

Another option is to plant alternative host plants when you plant the parsley. Seeds for fennel, dill, rue, and carrots should be at most stores that carry gardening supplies. Since rue won't be at a grocery, I encourage you to plant rue at the same time as the parsley. Rue is the alternative host plant that may provide the most interesting experimental results.

How Much Parsley Do You Need?

We plant bedding plants of parsley every spring. When it's mature, four plants of curly parsley (or Italian parsley) can bush out and cover an area 2 feet by 2 feet (.6 m x .6 m)—if they get plenty of water and sun. We may have as many as ten EBS caterpillars at a time on that much parsley.

Fig. 3.3 A large eastern black swallowtail caterpillar eating curly parsley. This one is in its fifth instar, almost ready to form a chrysalis.

You Can Grow Native Host Plants

If you like the idea of creating a lesson on native plants versus introduced and invasive species, this is a good time. None of the plants I've mentioned so far is native to North America, nor is Queen Anne's lace, which is common meadow plant in the United States that EBSs will eat (yes, it's in the carrot family). So what did EBSs eat before all of these introduced and garden species were carried here?

I know of three plants native to North America that they'll eat:

1. *Zizia aurea* (golden Alexanders)
2. *Polytaenia texana* (Texas prairie parsley)
3. *Polytaenia nutallii* (Nuttall's prairie parsley)

Where can you get these? Search the Internet for "Native plants North Carolina," except substitute the name of your own state. If your state has a native plant supplier, you can find it that way. I did that and learned that golden Alexanders is the only one of the three listed above that grows in North Carolina. So I could be planting that! That would be fun… and a great lesson about both the value of preserving native plants and the damage that invasive plant species do by displacing native plants. Local fauna usually aren't able to eat invasive plants that didn't evolve with them, so invasive plants grow out of control; whole ecosystems can be extinguished that way. A botanist friend of mind surveyed the plant composition of a park near my home and found that 35 percent of the plants growing there are not native to the area. No wonder there are so few animals in that park. Removal of invasive plants to restore native ecosystems is a huge expense for county, state, and federal government departments that oversee natural resources. (See more about that in our book *Going Green*, Fulcrum Publishing 2008.)

Finding information you can use about native plants, introduced plants, and invasive plants is also easy on the Internet—here's one good source: *www.ecosystemgardening.com*.

Of course, this is not to say that planting vegetable and herb gardens is a bad thing. We do have to eat. But we can choose native plants when landscaping and thereby benefit local birds and other wildlife.

Nectar Flowers Attract Adults

You can help attract adult EBSs to lay eggs on your parsley by planting nectar flowers to provide food for them. A caterpillar's job is to eat voluminous amounts of foliage for rapid growth to metamorphosis, but adult swallowtails eat only nectar to fuel their brief adult activities. The adult role of these butterflies is focused on reproduction: find enough nectar plants to survive, find a mate, mate, find host plants to lay fertilized eggs on. If they can do it all in your garden, that's a big bonus.

Not just any nectar flowers will do, however. A tube-shaped nectar flower, such as trumpet vine, is great for hummingbirds because hovering is required to reach the nectar. But butterflies must land to get at the nectar, so tubes are no good. Suitable nectar flowers that butterflies can land on easily include purple coneflower, milkweed, thistle, red clover, lantana, penta, ironweed, joe-pye weed, marigold, zinnia, impatiens, butterfly bush, calendula, and bachelor's buttons. Find more by searching the Internet, using the words "nectar flowers black swallowtails" or "butterfly gardening" or something similar.

When looking for nectar from the air, it is easier for butterflies to see colorful flowers grown close together.

If you can, it's nice to plant flowers that are indigenous (native) to your area, but not essential. Some nonnative plants can spread and become invasive, so keep them confined to your planting bed or garden.

What Happens to Butterflies and Caterpillars in Winter?

Black swallowtails go through several generations during a single summer, since one generation (egg > caterpillar > pupa > adult > egg) can happen in three to four weeks. Some of the eggs laid in late summer will grow to metamorphosis in September or October and will overwinter as pupae—inside the chrysalis. The adult and caterpillar stages can't survive cold weather, but the pupa inside the protective chrysalis can. In March, adults emerge from the overwintering chrysalises ready to look for nectar, for mates, and for host plants to lay their eggs on.

How Do You Find Caterpillars on Parsley?

I have a friend who can find swallowtail eggs on parsley, but I can't—or I don't look hard enough. One egg is about twice as big as a grain of salt, and they are deposited singly on the underside of leaves. I find the larvae more easily. Very young larvae (caterpillars) are dark brown or blackish with a white saddle mark; some say they're camouflaged as bird droppings. But their patterning changes every time they shed their skins (the caterpillar has five "instars," or stages), and these color changes are variable between individuals. Some stay largely black (with stripes) through several instars; others become white with black stripes and yellow spots. The larger caterpillars I've had are leaf green, with black stripes and yellow spots—the fifth instar is always these colors. They're beautiful! Although the colors and patterns provide camouflage, EBS caterpillars are preyed upon by birds, wasps (which may carry them away to the nest), praying mantises, and other insects.

The Caterpillars' Weapon

EBSs do have at least one method of defense beyond camouflage, although I'm not sure how effective it is. If you try to pick up a largish EBS caterpillar, you may see it poke out a fleshy, forked, yellow-orange structure from just behind the head. The structure, called an osmeterium, protrudes only about 3 millimeters. Simultaneously, the caterpillar thrashes around as though trying to wipe the thing on you (always unsuccessful at that, in my experience). A few seconds later, you may notice an odd, mildly stinky smell in the air. The osmeterium emits secretions containing foul-smelling terpenes, but the odor wafts away in seconds. When you put the caterpillar down, the osmeterium is drawn back inside its little groove. Normally the osmeterium stays hidden, but any threat can cause it to inflate with blood and suddenly evert. I don't think it deters predators very well, because caterpillars left outside usually disappear. That's why I always bring them inside if I need them for students.

How to Keep EBS Caterpillars Indoors

When you bring caterpillars inside, you must have host plants to put them on. You can either cut parsley stems outside or buy organic parsley at the grocery (remember, nonorganic has insecticides on it and butterflies are insects) to feed them inside. You can put the parsley in a vase with a small opening, or in a covered jar, poking holes for the stems through the lid or a through a foil cover. Caterpillars may fall into exposed water and drown—I've seen it happen. Make sure you put water in the container before putting the stems in. One bundle of organic parsley kept in a plastic bag in the refrigerator will last a week—you'll only need only a few stems at a time.

In my experience, EBSs will not leave the parsley as long as it's fresh, so the parsley doesn't need to be inside an insect sleeve cage or a box/bucket/terrarium covered with plastic wrap or Plexiglas. If you have it at school, though, you may need to keep it under wraps just to keep hands off the caterpillars—your call on that. Having it in the open definitely facilitates student observation.

You must check the parsley and replenish the water level daily. If the water gets cloudy or smelly, or if there's any rotting plant matter, change the water and remove

the decomposing stuff ASAP. If the parsley turns yellow or goes limp, change it for fresh parsley—caterpillars can't wait a day for edible food. I'm not sure exactly how long they can wait, but I do know it's not very long. They will pause for short periods after eating and sit immobile while their food digests. They also stop eating and sit still just prior to shedding their skins as they grow. They eat the skins they've shed, although they do it fast, so it's difficult to observe.

Fig. 3.4 *This chrysalis formed on a stick brought in from outside.* Photo by Ken Kneidel.

Time to Form a Chrysalis

When the caterpillars get close to 2 inches (5 cm) long, you know they'll be pupating (forming the chrysalis) soon (see Figure 3.3, a fifth instar caterpillar close to pupation). The time from hatching to metamorphosis depends on the temperature and volume of food provided steadily, but I believe it can be as short as nine days, although it's usually a bit longer. The caterpillars stop eating and will leave the parsley or other host plant, but before this happens, you need to put the parsley container inside a terrarium or a bucket/box with some kind of covering. Plastic wrap is fine as long as you loosen it occasionally to let air in. I use an insect sleeve cage (see page 24) with a mesh top that allows air to circulate and wedge a few bare sticks in the cage horizontally or angled (¼ inch wide is fine). I lean some of them up against the side to make them diagonal and usually tape the sticks in place at either end, making sure no sticky surface is exposed to little caterpillar legs. When a fully grown caterpillar leaves the parsley, it will wander around the sleeve cage looking for somewhere to transform into a chrysalis. Sometimes they pick a horizontal stick, sometimes vertical, sometimes diagonal, but it needs to be a place that will provide room for the new wings of the butterfly that will emerge from the chrysalis. Keep an eye on the caterpillar as it wanders around. If it stops moving on a stick, especially the underside of a horizontal stick, it's about ready to do its thing. It gets very still and empties its digestive system, creating a small puddle of goopy green stuff.

Over the next few hours the caterpillar grows more compact and forms a slight arc, with the middle of its body pulling away from the stick. At some point it attaches itself to the stick with two silk threads. Then, usually overnight, it suddenly begins to wiggle out of its caterpillar skin by splitting the skin on the head, like a human shimmying out of tight dress that winds up at her feet. Underneath is a smooth casing—the covering of the chrysalis, or pupa. It's amazing to see the caterpillar vanish as just a "skin." Imagine a human wiggling out of its skin, to reveal just a smooth oblong object underneath. It's very strange! When the skin falls away, the chrysalis remains where it is, suspended by the two threads.

Fig. 3.5 This adult eastern black swallowtail emerged from the chrysalis that's just behind it. The butterfly has been hanging upside from the stick for 30 minutes while its wings unfurl and dry. The chrysalis skin is empty now. Photo by Ken Kneidel.

It's important to leave the chrysalis alone; avoid jostling or touching it. If the threads are broken, then it will almost certainly die. After a couple of weeks, the skin of the chrysalis will break open, and the beautiful butterfly will crawl out, with its wings all wadded up. It will cling with its legs to the chrysalis shell or the stick for some time—maybe an hour or two—while pumping

blood into the arteries of the wings to extend them. In my experience, they usually cling to the underside of the stick so the wings are hanging down while this happens. Slowly the wings will unfold and straighten. Then it needs to let the wings dry. During the few hours after its emergence, the butterfly is extremely vulnerable. If the wings are touching anything in the cage as they unfurl, they will harden misshapen. If students handle it, they may knock scales off the wings and damage them. The butterfly needs to be undisturbed until it begins to crawl around of its own volition, with straight, functional wings. Then, if you have a butterfly that is native to your area, you can take it outside. It will stand on students' fingers for a while, making them very happy! This should be supervised to avoid accidents. In my experience, the period when the wings are "done" and the butterfly will agree to stand on fingers lasts only about ten to fifteen minutes. Then in one magical moment, off it flies! If you have nectar flowers, it's nice to let the butterfly go on them, although often it won't stay there—it just wants to get away from the activity.

If it's raining or very windy when the butterfly is ready to be released, you can offer it white grape juice, apple juice, or sugar water on a paper towel until the weather improves. But let it go as soon as you can, especially if it won't eat. More than twenty-four hours without feeding may affect its life expectancy.

This is an experience the students will never forget—and neither will you!

Eastern Black Swallowtails at School

Observations and Activities

Students can help with the planting or maintenance of the host plants or nectar flowers. Even if you've already grown an adequate supply of both, students can still contribute more.

As a student activity, consider providing a "station" with a bucket of soil, a roll of masking tape or labels, plastic cups, a waterproof marker, and bowls of host-plant seeds. Students can come to the station in pairs, add soil to their cups, and choose one seed type to plant in their cups. Tiny seeds (e.g., parsley) should be laid on top of the soil, not covered. Each student can use a marker to write his or her name and seed type on a piece of tape or directly on the cup. Place all the cups on a few cafeteria trays in a sunny window and let students take charge of watering them sparingly over the next few weeks.

If you have parsley growing outside, students can help monitor the plants for caterpillars every day. Small caterpillars can be moved indoors by pinching off a small sprig of the outdoor parsley without touching the caterpillar. You or the student can then place the small sprig onto mature parsley you have indoors, cut or potted. (Cover any open water in a jar or vase first.) The caterpillar can then move itself onto the bigger plant.

If you have enough caterpillars indoors, divide the students into groups and give each group a caterpillar on a cutting of parsley in a vase or jar. Ask if they have any questions. Then get them to observe carefully by asking them a series of questions. If you have only one or two caterpillars, you can write out the questions and let the students individually—in sequence—observe the caterpillar(s) and write their answers to the questions, while others are working on something else.

As the caterpillars grow, you might want to have each student keep a journal for ongoing observations and questions that occur to them. You won't have trouble getting the students to watch and keep track of the caterpillars, but they may need encouragement to write down what they notice. You can also keep a class log next to the caterpillars for students to write questions and observations that others can read.

Fig. 3.6 After we raised the caterpillar, Nia is ready to put the newly emerged eastern black swallowtail on a nectar flower in the garden and wish it well!

Questions for Students

Following are questions to ask students, either orally or in writing. Words in italics are my comments to you, the teacher.

Describe the colors and patterns you see on the caterpillar. Which is the background color? (If asking students questions orally, encourage them to say more than just "green," "black," "yellow," or "white." Try to draw out descriptions such as "stripes" and "spots." Asking for a description forces students to look carefully at the caterpillar.)

- Are the caterpillars cylindrical, like a paper-towel roll? *(No, the head end is bigger than the other end.)*

- How can you tell which end is the head? *(The head end is bigger.)*

- Do they have eyes? *(Not visibly so.)*

- Do they have antennae or soft "tentacles" on the head? *(No, but caterpillars of some other species do.)*

- Do they have horns or spines on the head or tail? *(No, but some caterpillars do. Tomato hornworms have a spine on the tail end; hickory horned devils have wild-looking spines on the head.)*

- Do they have spines or fuzz or any kind of surface covering? *(Earlier instars have short blunt spikes, but the fifth instar does not.)*

- What kind of appendages do they have for moving? *(Behind the head area, caterpillars have three pairs of jointed legs with hooks. These "thoracic legs" are the forerunners of the six legs each butterfly will have on its thorax as an adult. Behind the thoracic legs are five pairs of "prolegs." These are stumpy legs that are good at grasping stems. The prolegs disappear in the adult.)*

- How is a black swallowtail caterpillar different from a snake or an earthworm?

- How is it different from other insect larvae the students might have seen, such as mealworms or fruit fly larvae? How is it similar to other insect larvae?

- How are the caterpillars different from a mammal such as a dog?

- How are the caterpillars similar to an animal such as a dog? *(They're alive, they move around, they have legs, they eat, they poop, they have a finite life span, and they have a mouth, eyes, and a head of sorts. They react negatively to dangerous stimuli, such as being mildly squeezed. They try to defend themselves. They have mothers and fathers; they will probably mate and have offspring if all goes well. They have a heart and circulatory system, they have a nervous system, they have muscles, they have a digestive system, they breathe. They have a youthful period and an adulthood.)*

- Do they run from people like many animals do? *(No. They do react negatively to being picked up, but seem to have no objection to resting on a person's hand or arm, although they will soon begin looking for the host plant again.)*

Experiments for Eastern Black Swallowtails

Some of these experiments are designed to explore the EBS's niche or its relationship with its native environment. These experiments highlight the characteristics of the EBS that help it survive in the wild. Some of the experiments are designed to provide a fun way to test hypotheses.

Remember that the hypotheses I give are just examples. Most are written in the "if … then" format, but they don't have to be. Your hypotheses will be the predictions made by the class or a particular student. Your result for each experiment will be a statement of how your animals reacted to your experimental setup. Your conclusion is a statement of whether your prediction was confirmed or not. For each experiment, adding replicates increases your confidence in the validity of your conclusion, but they may be omitted if tedious for young children.

Experiment 1

Question:
Will eastern black swallowtail (EBS) caterpillars accept a new host plant in the same plant family?

Hypothesis:
We think EBS caterpillars will (or won't) accept a new host plant in the same plant family.

Methods:
Your EBS caterpillars are presumably feeding on the host plant they hatched on. For this experiment, you'll move at least one caterpillar to a related host plant. The carrot family (Apiaceae) is the only plant family I know of with several species that are host plants to EBS caterpillars, so try to do this experiment with two plants in the carrot family: parsley, dill, fennel, green carrot tops, or Queen Anne's lace. If your caterpillar is eating parsley, set up a narrow-mouthed vase or bottle with sprigs of fennel or green carrot tops. Or a potted plant is fine. Be sure the plants haven't been sprayed with insecticides; buying organic carrot tops will take care of that. As mentioned earlier, if you don't have a narrow-mouthed container, cover your container with foil and poke holes for the stems to eliminate the possibility of a caterpillar falling in the water.

It's best to transfer the caterpillar from one plant to the other without touching it, if possible. To do that, pull out one sprig from your original host plant that has a caterpillar on it. Cut away all the foliage on that piece except for the stem and the branch the caterpillar is sitting on. Put the trimmed stem in the container with the alternative host plant, so the caterpillar can easily reach the alternative plant.

If your caterpillar is on a host plant that's rooted outdoors, and your alternative host plant is also rooted, you can cut a small section of the original host plant with the caterpillar aboard and lay it in a secure place on the new host plant, or you can tie it on with a piece of thread, string, or a twist tie. Be wary of glue or tape that could trap the caterpillar.

Have students observe the caterpillar every twenty minutes for an hour. If it hasn't left the original plant after an hour, you can prod it gently to move or carefully lift it and move it. If a student moves the caterpillar, make sure he or she doesn't squeeze it. Have students continue checking every twenty minutes for another hour to see whether it's eating. If not, you can try a third hour, but if the caterpillar still isn't eating, return it to its original host plant. A prolonged period without food can damage a caterpillar.

Result:
Students' results are a statement of their own observations. The caterpillar will probably finish any remaining foliage on its original stem and may begin eating the stem itself, if the stem is soft enough. But in my experience, when the original host plant is gone, most EBS caterpillars will

eventually accept the alternative host plant and resume eating.

Conclusion:

Students accept or reject their hypotheses, based on their observations. EBS caterpillars can transfer from parsley to fennel, carrot tops, or another variety of parsley. Dill is sometimes rejected. Overall, EBSs seem to prefer more of their original host plant to switching.

Experiment 2

How caterpillars identify host plants: How can EBSs tell the difference between two plant types? The sharp, strong jaws caterpillars use to cut leaves are called maxillae. Sensory cells on the maxillae can taste whether a plant is a host plant. Caterpillars also have tiny antennae on the front of the head near the jaws that can detect volatile oils and other scents emanating from plants. These scents picked up by the antennae help in recognizing host plants.

Question:

Will EBS caterpillars accept a new host plant in a different plant family?

Hypothesis:

We think EBS caterpillars will (or won't) accept a new host plant in a different plant family.

Method:

As in the experiment above, you'll be moving at least one caterpillar from its original host plant to a different host plant: from carrot family to rue, or vice versa. Based on my experience, this procedure is most easily carried out by an adult, but students can monitor the results.

If your caterpillar is eating parsley or another plant from the carrot family, set up a narrow-mouthed vase or bottle with cut stems of common rue, available at many nurseries. Be sure the plants haven't been sprayed with insecticides (call and ask the nursery before you go). As mentioned earlier, if you don't have a narrow-mouthed vase or bottle, cover a cup or jar with foil and poke holes for the stems, to keep caterpillars from drowning.

It's best to transfer the caterpillar from one plant to the other without touching it, if possible. Over

Fig 3.7 To move a caterpillar to a new host plant, cut the stem of the plant the caterpillar is resting on and put the stem into the vase/jar containing cuttings of the new plant.

time I've noticed that frequently handled caterpillars are less likely to make it through complete metamorphosis successfully (they may never emerge from the chrysalis). To move the caterpillar without handling it, pull out one stem bearing a caterpillar from your original container. Cut away all the foliage on that sprig except for the stem and a minimum portion of the branch that the caterpillar is on. Put the trimmed stem in the container with the rue, so the caterpillar can easily reach the rue. All the cut stems should be in water.

If your caterpillar hatched on and has been eating rue, then do the opposite—transfer it to organic parsley, fennel, or carrot tops.

If your plants are rooted, you can tie the transferred cutting (bearing the caterpillar) on to the new host plant with string, thread, or a twist tie.

Have the students observe the caterpillar every twenty minutes for an hour. If it hasn't left the original plant after an hour, you can prod it gently to move or carefully lift it and move it. If a student moves it, make sure he or she doesn't squeeze the little critter. Students can continue to check every twenty minutes for another hour, and note whether it's eating. If not, you can try a third hour, but if the caterpillar still isn't eating, return it to its original host plant. A prolonged period without food can damage a caterpillar.

Result:

Students' results are a statement of their own observations. In my experience, EBS caterpillars from a carrot-family plant will not eat rue, even if I physically place them on it and they have no choice. I do know one person who raises EBSs and can transfer hers successfully, however—maybe she waits longer than I do.

I have more ambiguous results with moving caterpillars from rue to parsley. I've transferred four EBS caterpillars from rue to parsley, caterpillars ranging from ½ inch to nearly 2 inches (1.3–5 cm) in length. All refused to eat for at least three hours, maybe more. The three smallest ones did begin to eat the parsley at some point, but did not continue eating it for long and never did make much fras (caterpillar feces), so they didn't appear to be digesting it well. The largest one never took a single bite of parsley. I moved them all back to rue, but they then seemed reluctant to eat the rue! I worried about them and wondered if I'd ruined their appetites permanently. After a day of nibbling, though, they did pick up the pace and resumed eating the rue with gusto. In the end, they all successfully metamorphosed into butterflies.

Conclusion:

Students accept or reject their hypotheses, based on their observations. In my experience, EBS caterpillars do not easily transfer from rue to carrot family, or vice versa—and perhaps not at all. My own observations suggest that the taste and scents of the original host plant or plant family may somehow imprint a preference on a caterpillar during the first or second instar of larval life.

Experiment 3

Question:

Does temperature affect the growth rate of caterpillars?

Hypothesis:

We think a small temperature difference will not affect their growth rate.

Method:

To carry out this experiment, you'll need at least two small EBS caterpillars of the same size. Or if you have access to caterpillars of another species that are the same size, any two caterpillars of the same species and same size will do. Locate two areas in your classroom that differ from each other consistently by two to three degrees in temperature. (Not in front of an AC vent—too much air current.) Since heat in a room rises, one spot could be on top of your highest shelf in an area close to a sunny window (but not in direct sunlight) or on top of a refrigerator. A cooler spot might be on or close to the floor in an area away from a window, or in a basement or a crawl space that doesn't get close to freezing temperatures.

Prepare two narrow-mouthed vases or jars with stems of the host plant in water. If you use jars, cover the mouth of the jar and poke holes in the covering for the stems. Put at least one caterpillar on the plants in each container. Put the containers in their respective locations.

Check the water level for the plants every day and make sure the plants are still fresh. Every day try to measure the length of each caterpillar without touching it. Caterpillars often stretch out on a stem between bouts of eating, making measurements easy, or, you may be able to straighten out its body by maneuvering the stem. If not, just try again later. But the measurements for the two caterpillars must be done at roughly the same time in order to compare their growth. It doesn't matter if the intervals between measurements are entirely regular (e.g., skipping a day for both locations isn't a problem).

Measure each for at least a week, or until the caterpillars look like they're almost big enough to form a chrysalis (1½ or 1¾ inches long [3.8 cm or 4.5 cm]). At that length, an EBS caterpillar is ready or almost ready to pupate (form a chrysalis). To do so, the caterpillars must have access to a safe surface to attach themselves to, so you'll have to put them back into their terrarium or sleeve cage.

After returning them to a covered cage or terrarium, keep watch daily and record the date each becomes a chrysalis.

Result:

Students' results are a statement of their own observations. The caterpillar in the cooler area should have a somewhat slower growth rate and hence shorter length at the end of the experiment. The "cooler" caterpillar will probably pupate after the caterpillar in the warmer area. The difference could be anywhere from one day to one week.

Conclusion:

Students accept or reject their hypotheses, based on their observations. The growth rate of poikilothermic, or "cold-blooded" animals such as caterpillars, depends on the ambient temperature. The lower the temperature, the slower the growth rate, within lethal limits. Which temperature would be best for caterpillars outside, the warmer or the cooler? Caterpillars are very vulnerable to predation. So the shorter the larval stage, the better the chances are of surviving long enough to metamorphose and reproduce.

Experiment 4

Question:

Since EBS caterpillars have legs specialized for grasping twigs, can they also walk on a flat surface?

Hypothesis:

We think they can't walk on a flat surface.

Method:

Before stating a hypothesis, have the students observe how the caterpillars move while traveling down a stem of the host plant. The first three pairs of legs are jointed, with hooks on the end for grasping plants. Behind the thoracic legs are two segments with no legs and then four pairs of fleshy unjointed "prolegs" that are good at clamping on to stems. The prolegs are followed by another segment without legs. On the last body segment, at the hind end, is one pair of anal prolegs. All the prolegs disappear in the adult butterfly, leaving only the three pairs of thoracic legs (which of course look different on butterflies). A caterpillar moves forward in a rippling way, contracting body muscles in much the same way as an earthworm.

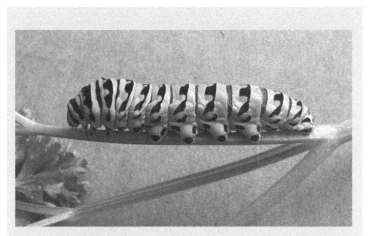

Fig 3.8 *The three tiny pairs of legs on the left, just behind the head, are jointed thoracic legs or true legs, with hooks on the end. The fleshy pairs of legs on the right are prolegs. Both are good at holding stems.*

While clutching the stem with its prolegs, a caterpillar contracts the muscles in the rear of the body, constricting that end and forcing blood into the front of the body. This extends the front of the body. The caterpillar then grabs the stem with the hooks on its thoracic legs and pulls the rear part of the body forward. Each leg moves in unison with its paired leg on the other side of the body.

After the students have observed that the caterpillars move by grasping stems, ask them to predict or hypothesize whether the leg and proleg action they've observed will work on a flat surface.

To accept or reject their hypotheses, students can observe in small groups how a caterpillar walks.

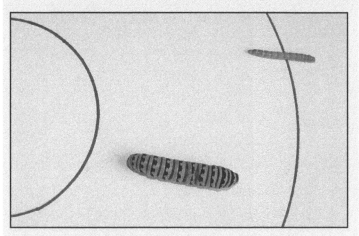

Fig 3.9 *Even though the legs are ideal for grasping stems, a mature black swallowtail caterpillar can walk very well on a flat surface, as on this laminated racetrack. Although this one is losing the race to a mealworm, mature EBS caterpillars can cover the distance in seconds. They have to be able to walk on flat ground in order to find a good place to pupate (form the chrysalis), if the host plant stems aren't sturdy enough.*

For the sake of the caterpillar, either you or a very careful student should place the caterpillar on a flat tabletop, with students sitting in a perimeter around the caterpillar. The floor doesn't work as well because the legs aren't as visible.

Result:

Students' results are a statement of their own observations. When placed on a flat surface, an EBS caterpillar will tap the surface with its face using the two tiny chemosensory antennae on the front of the head (these antennae are so small that you probably won't see them). It may do this for up to five minutes. Then the caterpillar will usually start walking. It uses the same method of forward motion on a flat surface as on a twig. The only difference is that the legs are not pinching a stem. An EBS caterpillar can move surprisingly fast on a flat surface (see Figure 3.9).

Conclusion:

Students accept or reject their hypotheses, based on their observations. It makes sense that caterpillars should be able to walk on a flat surface. In nature, a fully grown caterpillar usually leaves the host plant to find a twig or durable surface that will survive the winter before making its chrysalis. This requires taking a hike across the ground.

Experiment 5

Question:

Do EBS caterpillars have preferences regarding where to make their chrysalises (natural twigs, artificial surfaces, slick surfaces, rougher-textured surfaces such as fabric; vertical, horizontal, diagonal)?

Hypothesis:

We think the caterpillars will prefer to make their chrysalises on a natural surface, such as a twig.

Methods:

This experiment really requires a number of caterpillars to answer the question, but the students like to make a hypothesis even if you have only one. Place a variety of substrates in the terrarium with the caterpillar(s). Get a multibranched stick outside, with some of its branches ¼ to ½ inch (.6–1.3 cm) in diameter. Wedge it into the terrarium so that it doesn't rock, and place it in such a way that

some branches are horizontal, some vertical, and some diagonal. You may want to tape one end of it to the wall to keep it steady, covering all the sticky surfaces of the tape. Offer a swatch of fabric by draping it over the top or taping it to the wall of the terrarium. A rougher fabric is better—but not terry cloth, which can catch their legs. The side of the terrarium will provide another option where caterpillars may make a chrysalis.

Fig 3.10 The EBS caterpillar that formed this chrysalis chose a chopstick. Note the two silk threads attaching it to the chopstick.

Result:

Students' results are a statement of their own observations. In my experience, most EBS caterpillars choose twigs as the place to make chrysalises. When they are fully grown caterpillars and ready to pupate, they usually leave the host plant and wander around the terrarium looking for a spot they like. Of the eleven EBS chrysalises I have as of this writing, seven are on dead woody sticks that I positioned in the terrarium. One chrysalis is on a piece of fabric, one is on the slick plastic side of the terrarium, one is on a stemlike metal rod, and one is on a chopstick. All of the sticks they're using are ¼ to ½ inch (.6–1 cm) in diameter, except one that's a very thin parsley stem, less than $^1/_8$ (.3 cm) inch in diameter. Eight of my eleven chrysalises assumed a horizontal, belly-up position on the underside of the surface they've attached themselves to. Of the other three, one is vertical, one is diagonal, and one is between vertical and diagonal.

Conclusion:

Students accept or reject their own hypotheses and draw their own conclusions. In my experience, EBS caterpillars will use what they can find as a surface to form the chrysalis, but given choices, most attach themselves horizontally to the underside of a woody stick.

Suggested variation: Offer the caterpillars something that is roughly the shape of a twig or stick from outdoors but is a human-made object, such as plastic and wooden chopsticks, a section of a thin wooden dowel, or a wire coat hanger. My guess is that they will choose something made of unfinished wood because the rough texture makes it easy for them to attach their two silk lines.

Fig 3.11 Unlike the chrysalis in Fig. 3.10, this EBS chrysalis is attached to cloth, and is beige colored. EBS chrysalises can be green, beige, brown, or black.

Experiment 6

Question:
Do EBS caterpillars avoid light?

Hypothesis:
Since EBS caterpillars are not nocturnal, we think they do not avoid light.

Methods:
One way to test this is to put an EBS caterpillar in a T-maze. Unless you have multiple caterpillars and several very careful students, this is a good experiment to do as a class. For instructions on how to make a T-maze, see Experiment 3 of Chapter 5 on bessbugs. Situate the T-maze on desk where all the students can see it. Fasten a thickish cloth over a flashlight with a rubber band so that the light will not be extremely bright. Put the flashlight a couple of inches away from one of the exits of the T-maze, turned on and pointing toward the exit. To start the caterpillar in the maze,

pinch off a stem that the caterpillar is on and place it at the bottom entrance to the maze. Nudge the caterpillar gently to leave the stem. Don't squeeze the caterpillar in any way, and touch it as little as possible. The caterpillar may spend five minutes feeling around with its head, but it will eventually take off down the corridor of the T-maze. Which way does it turn? Toward the flashlight or away from it? If your caterpillar is willing, repeat this action up to ten times. You can restart the caterpillar by scooping it up with a piece of paper and moving it back to the maze entrance.

Result:
Students' results are a statement of their own observations. In my experience, the caterpillar has no preference for either direction in a T-maze, whether a light is present at one of the exits or not.

Conclusion:
Students draw their own conclusions. EBS caterpillars have no obviously visible eyes. Most caterpillars do have six pairs of very small and simple eyes called ocelli, which can detect a change in the intensity of light but do not form an image. I assume that EBS caterpillars have them too, although I can't see them. In my experience, EBS caterpillars do not respond to moderate changes in lighting or to movement, such as my hand moving in their terrarium. I have only one observation that suggests an ability to detect a change in light: One time, I shone a small pocket flashlight at a caterpillar eating on its host plant, moving the flashlight closer and closer as I continued to get no reaction. When the flashlight was a few inches from the caterpillar, it stopped eating. I turned off the light right away, and it resumed eating after a few minutes. That one observation doesn't prove anything, but it was food for thought.

Experiment 7

Question:
Do EBS caterpillars tend to stay hidden in the foliage of their host plant?

Hypothesis:
We think EBS caterpillars will stay hidden in foliage when possible.

Methods:

You need to have a fair amount of the host plant to test this—enough that it would be possible for a caterpillar to conceal itself. Have the students observe the caterpillars every five or ten minutes for an hour. During each observation, have them record whether each caterpillar has more than half of its body in plain view or more than half of its body hidden from view. Also have students record whether the caterpillars are in the top quarter of the plant or are lower.

Result:

Students state their own observations as their result. In my experience, caterpillars make no apparent effort to conceal themselves in foliage. Their ocelli could guide them to more shaded areas, where light is less intense. Perhaps outdoors, where sunlight can be bright, the caterpillars do avoid the very top of the plant where they would be most exposed to predators. Indoors, however, I have not observed them to do so.

Conclusion:

Students draw their own conclusions. Some species of caterpillars may hide, but my impression is that EBS caterpillars rely on camouflage and a bad smell to protect themselves from predators. The dark color of the young caterpillars, coupled with a white saddle mark, is thought to resemble a bird dropping. The green, black, and yellow colors and the stripes of an older EBS caterpillar blend in with green leaves, as the stripes of a zebra help conceal it in an area of tall grasses and shadows. In addition to camouflage, the EBS swallowtails will evert the osmeterium if threatened, releasing a stinky smell. The adult EBS mimics the coloration of the toxic adult pipevine swallowtail. Predators may mistake the adult EBS for the pipevine, which they have learned to avoid.

Experiment 8

Question:

Will EBS caterpillars seek cover if it's offered?

Hypothesis:

We think EBS caterpillars will seek cover when available.

Fig 3.12 EBS caterpillars seem unfazed by the light of day and lack of cover.

Method:

This is most easily done in groups or as a class experiment. To make a simple refuge or hiding place for a caterpillar, create a small horizontal shelf in a shoebox. Do this by wedging a piece of cardboard in place so that it rests uniformly about an inch above the shoebox floor (at one end of the shoebox) and covers about a quarter of the floor. This creates a small, dark place for the caterpillar to hide if it has an urge to. If students are to work in group, make several of these test boxes. Place an EBS caterpillar in each shoebox, and cover the box with clear plastic wrap or a see-through mesh, secured with a rubber band or tape. Have students check the caterpillar's location every five minutes for half an hour, or every ten minutes for an hour. Don't leave the caterpillar in the box for more than an hour, because it needs to eat often and fasting can damage it permanently.

Result:

Students' results are a statement of their own observations. If an EBS caterpillar has no urge to seek cover, and if it stays on the floor of the box, then we would expect to find the caterpillar under the shelf in one-quarter of the observations just by chance. If instead the EBS caterpillar is located under the shelf in half of the observations or more, that would suggest a preference for cover. In my experience, an EBS caterpillar does not seek cover.

Conclusion:

Students draw their own conclusions. In my experience, an EBS caterpillar does not seek cover in a shoebox. Since a caterpillar's role in a butterfly life cycle is to eat and grow as rapidly as possible, the most adaptive behavior when not on its host plant is to walk around in search of a plant. It may stop walking to rest or to conserve its energy, but in my experience, EBS caterpillars do not demonstrate an attraction to dark hiding places.

Experiment 9

Question:

Are EBS caterpillars positively thigmotaxic? That is, do they prefer locations that allow their bodies to be in contact with walls or a roof?

Hypothesis:

We think EBS caterpillars are positively thigmotaxic, or wall seeking.

Methods:

You can test this with a T-maze. Instructions for making the T-maze are described in Experiment 3 of Chapter 5 on bessbugs. If you have several T-mazes and several caterpillars, students can do this in groups, or you can do it as a class experiment. To get the caterpillar from its terrarium into the maze, pinch a stem that the caterpillar is on and place it at the bottom entrance to the maze. Nudge the caterpillar gently to leave the stem, lifting it gently if needed (no squeezing). The caterpillar may spend five minutes feeling around with its head but will eventually take off down the corridor

Fig 3.13 EBS caterpillars, in my experience, don't show any wall-seeking or wall-hugging tendencies, but move right down the center of a T-maze.

of the T-maze. As it walks down the corridor, have students observe whether it tends to hug one wall; that is, does it keep one side of its body in contact with one wall? If it hugs the left wall, it will most likely turn left at the top of the maze to maintain contact with the wall. If it hugs the right wall of the long corridor, then it will most likely turn right at the top of the maze. If your caterpillar is willing, repeat this trial up to ten times.

You can restart the caterpillar by gently scooping it up with a piece of paper and moving it back to the starting point.

Result:

Students' results are a statement of their own observations. In my experience, EBS caterpillars do not tend to hug one wall or the other of the corridor. So in a T-maze, the direction they turn at the top is random.

Conclusions:

Students draw their own conclusions. In my experience, EBS caterpillars have no urge to stay in contact with a wall. It makes sense, because they have no opportunity to hug walls or seek confined spaces in nature.

Additional Resources for Black Swallowtails

General care:

Stout, Todd. "Eastern black swallowtail." *Raising Butterflies*, www.raisingbutterflies.org/eastern-black-swallowtail.

Natural history:

Hall, Donald. "Eastern black swallowtail: *Papilio polyxenes asterius* (Stoll) (Insecta: Lepidoptera: Papilionidae)." University of Florida, IFAS Extension, 2012, edis.ifas.ufl.edu/pdffiles/IN/IN90600.pdf.

Videos:

"Black swallowtail butterfly into chrysalis." YouTube, www.youtube.com/watch?v=Zc4w8VuYqE4.

Motleypixel. "Black swallowtail caterpillars pupate." Vimeo, vimeo.com/15930001.

Chapter 4
Mystical, Magical Monarchs Need Our Help

A Charismatic Critter

Monarch butterflies are among the most beloved of all insects. I've heard them referred to as the "Bambi of the insect world," and I understand the sentiment behind that phrase. In California, monarchs are celebrated in annual festivals! I've seen their images festooned across buses here in North Carolina. What other real insect is featured in an IMAX movie? Just yesterday, I saw fabric monarchs for sale in a gift-wrap department—the only animal represented there. Monarchs are famous! And rightly so. They have a remarkable and unique life history—these butterflies migrate every year farther than many migratory birds do, and farther than any other insect species. The migration takes them from the United States and southern Canada to their wintering grounds in the highlands of central Mexico. And then back again in spring. An adult monarch weighs only ¼ ounce or ½ gram— the weight of a piece of computer paper. How do they fly so far? And how do they know where to go? No one really knows. The mystery is part of their allure.

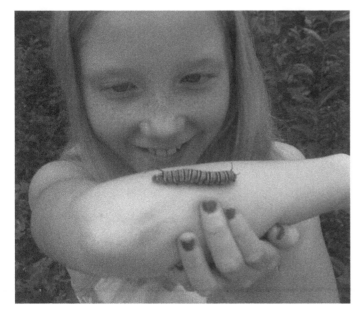

Fig. 4.1 Kids love monarchs!

Monarchs are popular, but unfortunately their survival is threatened. Because they are both beloved and in danger, monarchs are ideal subjects for engaging your students personally in hands-on conservation. Through the tools in this chapter, your students can enter the world of "citizen science," can contribute to ongoing monarch-conservation research, and can interact with students whose lives are radically different from their own. Nothing can be more empowering to your students than seeing themselves as part of a solution. With the resources in this chapter, that can happen. Teachers say they've seen their students' worldview and global perspective changed by their involvement in the community of monarch conservation. Our fragile planet needs your students to grow up aware and empowered to take on the challenges of our already overpopulated and overharvested planet. This is the richest opportunity I know of for truly informing and inspiring young students about ecological activism and its global aspects.

So, in what follows, I'll describe the best of the monarch science organizations—those that teachers can easily plug into. The most effective way to show you what can be done is to introduce you to a sixth-grade teacher, Jennifer Joyner, who has gotten her whole school involved in monarch conservation. She (and her coworkers) have done it all with the resources described to you in this chapter. I'll review everything Jennifer did to get to where she is, and how her students have responded.

Before we get to Jennifer and her school, though, let me get you better acquainted with this remarkable butterfly, and the threats to its survival.

Recognizing a Monarch Butterfly

Monarchs range over most of the United States. You could come across a monarch in any season other than the cold of winter. You might see one on a nectar flower, or on milkweed, or just flying by. I saw one fly through my yard a couple of weeks ago in September. It stopped on a marigold for a few moments, then flew off across my neighbor's yard. If you live in a migratory route, you might see large numbers of them fly by. How would you recognize a monarch?

Fig. 4.2 Monarch butterfly on milkweed. It's been drinking nectar from the milkweed flowers and has laid a few single eggs on the underside of leaves.

Adult monarchs are large orange butterflies with deep and powerful wing strokes. They often glide with the wings held in a V. The wings have black edges, with white spots on those edges. The wings also have conspicuous black veins throughout. You can easily find pictures of them on the Internet. Viceroy butterflies look very similar but have shallower wing strokes and sail with wings held flat rather than in a V. Viceroys are also a tad smaller than monarchs. Fritillaries are a large family of similar-sized butterflies that are often orange and, at a distance, can be mistaken for monarchs. But fritillaries are more likely to have black spots or other intricate coloration rather than the black edging with white spots and the back veins of monarchs.

Monarchs' Mind-Boggling Migration

Every spring millions of monarch butterflies leave their winter roosts on a few mountain peaks of central Mexico and fly 2,000 miles (3,200 km) to the Gulf Coast of the United States. Think for a minute about how odd that is—an animal so tiny and fragile flying so far. Monarchs are the only butterflies to fly such a long two-way migration every year. But here's the really crazy thing. The following autumn, the great-great-grandchildren of those monarchs return to the exact same sites in Mexico's

central highlands. These are individuals who've never been there, who have no one to lead them, and who have a brain smaller than a sesame seed. Their navigational cues remain the subject of much speculation.

Why do monarchs migrate? Like most insects and other poikilothermic (cold-blooded) animals, butterflies can't tolerate subfreezing conditions. Also, their foods aren't available in cold winters. Most other insects face these challenges by entering a dormant state similar to hibernation. For lots of insects, such as praying mantises, the dormant overwintering stage is the egg, which doesn't need to eat. Black swallowtails and most other butterflies overwinter as a pupa in the chrysalis; the pupa doesn't need to eat either.

Most butterfly species are not migratory, and monarchs weren't either originally. Thousands of years ago, they lived in Mexico year-round, eating milkweed. As the Ice Age ended, and the ice covering the United States retreated northward, milkweed began to spread northward. As the milkweed edged northward, monarchs followed it to exploit the additional food resource, laying their eggs on the northern milkweed in summers, but they flew back to Mexico for the warmer winters. As milkweed continued to progress northward all the way into southern Canada, the return to Mexico got farther and farther. But the monarchs kept doing it, and they still do.

Fig. 4.3 Milkweed is the required food (host plant) for monarch caterpillars. This full-size monarch caterpillar is busily chomping a leaf.

Monarchs from east of the Rocky Mountains now migrate to twelve sites with very specific habitat in the mountains west of Mexico City, where they overwinter. You can see a map of the sites in relation to Mexico City on this webpage: www.learner.org/jnorth/tm/monarch/MexicanOWSites1_2.html. The region encompassing all twelve of the sites is only 73 miles wide! For months the adult butterflies roost in fir trees at these sites, in a dormant state—not active and not needing any nectar. My friend Beth Henry visited one of these sites and said the trees were orange with monarchs! The climate there is cold enough to keep them dormant, but not cold enough to freeze them. And the continuously moist air keeps them from dehydrating.

During the second week of March, the monarchs begin to leave the fir trees and head north and east toward the southern United States. These adults have had a very long life for a butterfly—eight or nine months—long enough to fly south, overwinter, and then fly north again. They arrive in the southern states with tattered wings and lay eggs on newly sprouted milkweed. These adults soon die, but their eggs hatch, and in late spring or early summer, a new generation of monarchs spreads northward. To see a map of the routes they take, look at the website for MonarchWatch (www.monarchwatch.org/tagmig/spring.htm). This new generation, the offspring of the monarchs that made the long flight to Mexico, continue to fly northward. At sites across the eastern and midwestern United States and southern Canada, they mate and lay eggs on milkweed, continuing the cycle of life. The lifespan of summer generations of adult monarchs is only three to five weeks, much shorter than the overwintering generation. During the summer, the monarchs go through three or four complete generations, depending on the duration of warm weather at their particular northern latitude.

As autumn comes and the weather cools, the last generation to emerge as adults is different from the summer generations. They look the same, but they don't mate right away or lay eggs. Instead they begin flying south toward Mexico, often in huge butterfly clouds. They are the great-grandchildren or great-great-grandchildren of the adults that made the same journey south the previous year. Every time I write that, it makes the hairs stand up on my arms!

How do they know where to go? Obviously, they have some kind of inborn navigational system. Some birds can migrate by the stars; others detect magnetic fields. Some wasps have been shown to use visual landmarks. For now, there are lots of guesses, but monarch navigation remains a mystery. However, we do know a little about how they are able to fly so far.

How Do They Fly So Far?

Monarchs, like migrating birds, accumulate fat reserves for fuel before starting out. And like migratory birds, monarchs are very efficient fliers. On the journey, the butterflies expend as little energy as possible to increase their chances of completing the trip before running out of fat. Monarchs do stop to nectar along the way. Some researchers believe that migrating monarchs glide on the wind, not beating their wings, whenever possible. If the wind isn't blowing just the right way, they may make maneuvers to correct for it, or they may drop to the ground until the wind direction is more favorable. Researchers also believe that they take advantage of thermals to gain altitude. A thermal is a warm body of air spiraling upward. Glider pilots have reported seeing monarchs as high as three-quarters of a mile above the ground!

Wintering in the United States

Monarchs that spend their summers west of the Rocky Mountains migrate to California rather than Mexico for the winter. The adult butterflies travel up to 700 miles to roost in eucalyptus trees and Monterey pine trees along the California coast. About 200 roosting sites have been identified from San Francisco southward. Any one site may have dozens to tens of thousands of dormant adults, compared to millions at the few sites in Mexico. In spring, the adults disperse in different directions—unlike the mass exodus northward from Mexico.

Fig. 4.4 Living in a California coastal town that celebrates the local monarchs, Meredith has a couple of monarch costumes. These wings were a gift, but her other pair is homemade! She also grows milkweed and sees lots of monarchs in her yard. Photo courtesy of Meredith Burrus.

Monarchs are very much a part of local culture near their roosting sites in coastal towns such as Pacific Grove, Pismo Beach, Morro Bay, and San Luis Obispo. Shops sell big monarch wings as costumes; some communities have monarch festivals when the butterflies arrive in fall.

Competition is fierce for California coastal property, but both nongovernmental organizations and government efforts protect some of the sites from development.

In southern Florida and a few tropical areas there are monarch populations that don't migrate at all!

Here's the Challenge: Milkweed's in Trouble, So Monarchs Are Too

Sadly, the number of monarchs in the Mexican overwintering sites has been declining since the 1990s. For a long time no one knew why, but monarch scientists at midwestern universities now think they get it. The majority of monarchs lay their eggs on milkweed plants that grow in or around cultivated fields in the midwestern United States. Those farms used to provide high quality monarch habitat. For years milkweed on farms produced more caterpillars than the same amount of milkweed in residential areas or roadsides.

But all that changed when Monsanto Corporation developed genetically engineered crops to be resistant to Monsanto's popular herbicide (Roundup). Of course, we all know that herbicides are commonly sprayed on agricultural fields these days to kill unwanted "weeds." Before corn and soybeans were genetically altered to withstand Roundup, crop fields could only be sprayed before the crops emerged from the ground, or else the crops too would be killed by the herbicides. But now, with genetically modified crops that are unfazed by the herbicide Roundup, many agricultural fields can now be sprayed throughout the growing season. And they are. Ninety-four percent of our soybeans are now genetically engineered to be Roundup resistant; 72 percent of our corn is. Monsanto calls their genetically modified seeds "Roundup Ready." The upshot of Monsanto's Roundup story is that nearly all the milkweed that used to grow in and around corn and soybeans (our most abundant crops) is gone, killed off by the herbicide Roundup. With the loss of agricultural milkweed between 1991 and 2010, monarch egg production in the Midwest dropped by 81 percent. Milkweed has almost disappeared from farm fields.

More Challenges: Milkweed Yanked from Meadows, Roadsides

Farm fields slammed by Roundup are not the only problem. Milkweed takes another hit when meadows or fallow fields are developed for human use, a growing phenomenon as our population grows. Native plants such as milkweed are not often chosen for landscaping around new buildings—although they can and should be. Native plants benefit all wildlife. Roadside mowing is yet another common practice that destroys milkweed. Roadsides are actually prime milkweed habitat.

Possible solutions to these threats to monarchs? One theoretical solution is to halt the use of Roundup on agricultural fields. A nice thought, but something that seems unlikely to happen soon: Monsanto is a huge and powerful corporation with too much control over legislators. When asked about the monarch issue, Monsanto replies that more data is needed—in other words, "No comment." To read more about Monsanto's frightening chokehold on American agriculture and its alteration of our crop seeds, check out the website of the Center for Food Safety, or our previous book, *Veggie Revolution* (Fulcrum Publishing 2005), or the eye-opening book *Monsanto vs. US Farmers*, available through the Center for Food Safety.

Other more immediate options include the cessation of routine roadside mowing, and an increased national awareness of the need to conserve and plant milkweed.

You and Your Students Can Help!

One powerful solution within our grasp is for multitudes of students and citizens to plant milkweed on school and residential properties—a very easy thing to do. Business properties too! Since milkweed is a native plant, it's adapted to our climate, unlike introduced tropical plants that need to be watered constantly and replanted every year. Once a milkweed plant is growing, it can survive weather fluctuations.

Planting milkweed is also a valuable step (though not essential) in bringing monarchs into the classroom.

"Citizen Science" Connections, Workshops, and More Monarch Teaching Tools

Before I tell you specifics about using monarchs in the classroom, I'd like to describe four extraordinary monarch organizations and the resources available through their interactive websites. These four sites (following) are loaded with opportunities to involve your students in hands-on monarch conservation, to learn from other teachers, and to connect your students with Mexican students living near the monarch overwintering sites. You'll also find an abundance of monarch lesson plans and videos, migration maps, gardening instructions and supplies, and more. Perhaps most exciting are the pages where you and your students can post observations to share with the community of monarch researchers and educators. The four organizations are all different. Later in the chapter you'll find the detailed account of sixth-grade teacher Jennifer Joyner's experience using the teaching resources offered by three of these sites.

Monarch Watch
(www.monarchwatch.org)

Monarch Watch was founded and is directed by a monarch ecologist at the University of Kansas, Chip Taylor. The organization's stated purpose is monarch education, conservation, and research. Monarch Watch is well-known for its tagging program. You may have heard of tagging and recapturing birds or other animals to track their movements. Monarch Watch offers classroom kits for tagging monarchs! The kits are available to anyone who wants to participate in this broad effort to learn more about monarch migration. A monarch tag is a tiny coded disk, sticky on one side, which is applied to the wing of the butterfly. If the monarch is later recaptured, alive or dead, and the tag information is reported to Monarch Watch, then the migratory route of that particular individual is known. The data from these recaptures are also used to understand the influence of weather on the migration and survival rate of monarchs. Of course, to tag a monarch butterfly, you have to find or raise the butterfly. More about that later in this chapter. The home page of Monarch Watch features links to other pages on the site. Topics you can access include how to grow milkweed, how to raise monarchs at home or in the classroom, migration and tagging, classroom projects, research topics, monarch conservation, and how to create a Monarch Waystation at your school or home and have it certified (simple—grow milkweed and nectar flowers!).

Journey North
(www.learner.org/jnorth/monarch/)

Journey North is an incredibly useful website for classroom teachers. If you're not interested in tagging,

Journey North is (in my opinion) a more useful website for teachers. It offers lesson plans and sheets for student responses on a huge variety of subjects related to monarchs, making use of the up-to-date information and maps on the site. I believe you could fill the entire school year with monarch activities and lessons from this website if you wanted to. Best of all, it's easy to contribute data to the website; all you have to do is spot a monarch outside. You report the sighting just by clicking on "Sightings" at the top of the home page. So easy. Then if you click on "Maps" on the home page, you can find your own dot on the appropriate map, click on that dot, and your short report will pop up on the screen, with your date of sighting, name of teacher, name of school, location, and any comments you made. Cool! Students enjoy that.

A memorable and unique option with this website is the opportunity to participate in a Symbolic Migration, where your students create small paper monarchs and send them (via Journey North) to students in a Mexican school near the monarch overwintering sites. The profile later in this chapter of teacher Jennifer Joyner describes this process in detail. Read about it by searching the Internet for "Symbolic Migration Journey North" or go to www.learner.org/jnorth/sm/index.html. Your students can read descriptions of the Mexican students' daily lives, written by individual students (in English or Spanish) at www.learner.org/jnorth/tm/monarch/LifeSanctuaryRegion.html. This up-close and intimate view of households in a country with a very different economy can launch a social studies unit or fuel Spanish studies. There are great opportunities here for cross-curricular connections.

Another great feature of this website is the impressive number of videos of monarch caterpillars and butterflies, many of which are coordinated with lesson plans available on the site. So, for an inquiry lesson about, say, caterpillar locomotion, students can watch the close-up videos that show the two different kinds of legs on monarch caterpillars and how the legs move in relation to each other. Really good, really interesting, really useful. Give it a try: you'll find both a great conservation message and great educational value.

Journey North is a subset of www.learner.org, a website funded by the Annenberg Foundation. The website's home page states at the top that the organization (Annenberg Learner) exists to provide "Teacher professional development and classroom resources across the curriculum." It certainly provides that.

Monarch Larva Monitoring Project (www.mlmp.org)

Like Monarch Watch, this site is the brainchild of monarch scientists at a major university. The Monarch Larva Monitoring Project (MLMP), a citizen science project started in 1996, solicits participant observations of monarch eggs and caterpillars. While most observations come from established sites that are observed weekly, I like the fact that the site also accepts anecdotal observations from anyone who cares to report them. Also, the observations are used to potentially impact monarch conservation. The primary scientist behind the MLMP is Karen Oberhauser, one of the biggest names in monarch research. The website states that the Monarch Larva Monitoring Project (MLMP) "is a citizen science project involving volunteers from across the US and Canada in monarch research. It was developed by researchers at the University of Minnesota to collect long-term data on larval monarch populations and milkweed habitat... As an MLMP volunteer, your contributions will aid in conserving monarchs and their threatened migratory phenomenon, and advance our understanding of butterfly ecology in general."

There are two different ways you and your students can submit information to the MLMP website. One is to register on the website as an MLMP volunteer and agree to monitor a milkweed site near you on a weekly basis during the milkweed growing season. As a volunteer, you report (via the website) specific information about any monarch eggs or caterpillars seen on the milkweed. The other option, which is less time consuming, is to report any observations you might make of monarch eggs or caterpillars on milkweed, even if it's just one time.

The MLMP is a part of a large suite of educational and conservation activities offered by the University of Minnesota MonarchLab. As of this writing, the MonarchLab offers "schoolyard garden grants" to facilitate the planting of milkweed and the study of monarchs in schools, or any plants that will encourage student observations of nature in the schoolyard. The application for the grant is on the website. The site also offers a classroom visit by a "monarch scientist" or a teacher workshop to be held at your school. Another MonarchLab project is Driven to Discover, which uses citizen science as a springboard for youth-developed research projects.

I like their detailed information and resources on planting and raising milkweed outdoors and on raising monarchs indoors. Their Monarch Store sells a wide variety of monarch educational materials, including a *Monarchs and More* curriculum guide (in book form, on paper) for grades K–2, 3–6, or middle school. Five lessons for each grade category can be viewed on the website as a sampling of what's in the book.

Check out the website to see what more is offered.

Monarch Teacher Network (www.eirc.org/website/programs-services/global-connection/monarch-teacher-network/)

Like Journey North, the Monarch Teacher Network (MTN) was created to be a teacher resource. MTN offers the famously effective Monarch Summer Workshops, which are two-day workshops scattered all over the country. They're designed for classroom teachers, but anyone can attend. Upcoming workshops and registration are on the MTN webpages. Your school can sign up to host a workshop too. Jennifer Joyner credits the MTN workshop with launching her confidently into the world of using monarchs as teaching tools. Read more on Jennifer later in the chapter, where you'll see just how much monarch education has rocked her world and her classroom.

There's also a Monarch Teacher Network page on Facebook, where you can interact directly with other teachers experienced in bringing monarchs to their classrooms and in growing milkweed.

According to the MTN home page on the website, "The Monarch Teacher Network is a growing network of teachers and other people who use monarch butterflies to teach a variety of concepts and skills, including our growing connection with other nations and the need to be responsible stewards of the environment."

The MTN webpages are a subset of eirc.org; EIRC is the Educational Information and Resource Center.

Planting a Monarch Garden: Milkweed and Nectar Flowers

Planting milkweed and nectar flowers is a great thing to do, even if you go no further. It's a hands-on activity that can accompany a science unit on animals, plants, migration, ecosystems, food webs, habitats, or environmental issues. You can attract monarchs with fewer than ten milkweed plants if they're in full sun and blooming well. Colorful nectar flowers nearby help monarchs flying overhead to notice your milkweed. Planting milkweed is a proactive statement of support for a species suffering from habitat loss, a step that models environmental activism for your students.

Students can, of course, help prepare a sunny garden bed to receive the seeds or young plants, help put them in the ground, help water them, and later check them for monarch eggs and larvae. Full sun is important to make the plants grow well and bloom. I've seen beds with thirty plants and no monarchs because the plants are partially shaded. I've seen other beds with two to eight flourishing plants in full sun that have attracted monarchs the first year.

You have lots of options for getting seeds or plants. You and/or students can gather seeds in the fall from milkweed plants in your area. Most milkweed species grow especially well in disturbed areas, such as vacant lots, gardens, farm fields and pastures, (safe) roadsides, or any sunny area that isn't paved or mowed. It's quite possible there's milkweed in some neglected area of your schoolyard, or in a student's yard.

You can sometimes buy milkweed seeds or plants from local suppliers. Monarch Watch provides a state-by-state list of milkweed seed providers, although not every state is listed. Or you can order seeds from other vendors online, such as www.butterflyfarm.com or www.educationalscience.com.

If you've bought seeds, planting instructions should be on the packet. And if you acquire young plants, transplant them as you would any other garden plant. Water the soil around the roots thoroughly before removing the plant from the ground or container, and retain a moist dirt ball around the roots throughout the transfer, otherwise roots may be damaged by exposure to air. Once the plant is in its new location, water it thoroughly and firm the soil with your fingers to eliminate any air pockets around the roots. The plants will need plenty of sun to flourish.

For more detailed instructions on finding milkweed and growing specific varieties of milkweed, click on "Monarch Lab" on the MLMP home page, then on "Monarch Rearing" (www.monarchlab.org/Lab/Rearing/FindingMilkweed.aspx, www.monarchlab.org/Lab/Rearing/GrowingMilkweed.aspx).

If you want to cut right to the chase, you can order from Monarch Watch a Monarch Waystation Seed Kit that includes seeds for nectar flowers and milkweed, as well as a detailed Creating a Waystation guide. Click on "Butterfly Gardening" on their home page, and then click on "Order" in the left margin (or you can go to monarchwatch.org/bring-back-the-monarchs/resources/plant-seed-suppliers).

Bringing an Egg or Caterpillar Indoors to Raise

Finding and Recognizing a Monarch Egg or Caterpillar

A monarch egg is very tiny (1–2 mm high, 0.9 mm wide) and is pale yellow or off-white. To the naked eye it looks sort of barrel shaped. When magnified, its shape is more like an avocado with longitudinal ridges—flat on the bottom as though the end were sliced off the avocado. The eggs are deposited singly on the underside of young milkweed leaves, usually no more than one egg per plant.

Caterpillars are easier to find than eggs, and they get easier as they get bigger. But the earlier you find them, the longer you get to watch them develop. To help you identify caterpillars and recognize which of the six stages of their development you have, check out the following descriptions, each called an instar. When a caterpillar has grown and sheds its too-tight skin, the shedding marks the end of that instar and the beginning of the next.

- The first instar is the one that emerges from the egg. During this instar the hatchling grows from .08 to .24 inches (.2–.6 cm) in length. It's pale green or pale gray with a black head capsule. It stays on the underside of the milkweed leaf eating the fine leaf hairs, then it begins eating the surface of the leaf.
- The second instar grows from .24 to .35 inches (.6–.9 cm) and has black, yellow, and white bands. Both first and second instars have a yellow triangle on the head with yellow bands around the triangle.
- The third instar increases in length from .39 to .55 inches (1–1.4 cm). The black and yellow bands on the back two-thirds of the body grow darker and more pronounced. The yellow triangle on the head is gone.
- The fourth instar grows from .51 to .98 inches (1.3–2.5 cm) and acquires distinct bands along the entire body, with white spots on the legs. The black tentacles on the front and rear of the body (often mistaken for antennae) grow much longer. The tentacles have a sensory function. The true antennae are very short and are on the front of the face, as in black swallowtails.
- The fifth and last instar increases from .98 to 1.8 inches (2.5–4.5 cm) in length. The fleshy prolegs

on the abdomen are now much bigger than the jointed true legs closer to the head. The colors and patterns are becoming still more distinct and vivid. This fully grown caterpillar can walk much faster than earlier instars. It must be able to leave the milkweed and search for a good place to make its chrysalis.

Feeding and Housing the Caterpillars

To bring a caterpillar indoors, snip off the branch of the milkweed plant that's bearing the caterpillar, including several leaves. Tender young leaves are preferred. Place the branch in a vase or jar of water. If it's a jar, cover the opening with foil to keep the caterpillar from falling in and drowning.

A very young caterpillar is unlikely to leave the milkweed as long as you keep a supply of edible fresh leaves in the container. The caterpillar can transfer itself from an old stem to a new one if you make the leaves touch. No matter how gently you do it, lifting the caterpillar risks injuring it, so best to let it move itself.

When the caterpillar reaches the fifth instar, you can make a cage for it if by covering a wire tomato cage with mesh. Tomato cages are sold at gardening stores and gardening departments in hardware stores; they're designed to surround and support floppy tomato plants in gardens. I recommend mesh because you and the students can see through it. Something like tulle or fiberglass window

screening is ideal, but any breathable covering will do. If the holes are too big, of course, the caterpillar can crawl through.

Put a multibranched stick in the jar or vase holding the milkweed, inside the cage, to provide a suitable pupating surface for the caterpillar. Note that some of the branches must be horizontal.

If you don't have a cage, you can hope that the caterpillar will stay with the vase or jar and pupate on the stick with horizontal branches that you've put inside. Sometimes the caterpillar will cooperate and use the stick. Often, however, it won't, and it may very well set off on a long trek across the floor looking for a better place to pupate, in which case the caterpillar is quite likely to be stepped on—especially in a classroom. This is why a cage is useful for a classroom. For reasons I don't understand, monarch caterpillars are much more likely to wander before pupation than swallowtail caterpillars are.

Of course, another option is to set the vase, milkweed, caterpillar, and stick outside when the fifth instar gets close to pupation. This does risk predation, though, and you'll probably miss seeing the fascinating process of pupation!

Forming the Chrysalis

When the fifth instar has grown to about 1.8 inches (4.5 cm) in length, pupation is imminent. In my experience, a grown caterpillar prefers a horizontal—or nearly

Fig. 4.5 Monarch caterpillars in the fifth instar have long floppy sensory tentacles on the head and tail end. They almost look like puppy ears! This one is eating milkweed. Its fleshy prolegs are visible.

Fig. 4.6 Alley, a Charlotte Country Day School student, holds a monarch cage made from a tomato cage covered with mesh. The CCDS students use these cages for caterpillars and milkweed they bring in from the school garden. The sticks across the top of the cage are for the caterpillars to pupate on. Photo courtesy of Jennifer Joyner.

horizontal—surface to pupate on. A stick works well, but I've seen caterpillars attach themselves to all sorts of artificial surfaces. When it finds a horizontal surface that's satisfactory, the caterpillar will stop on the underside of the surface and just stay immobile for a while (see Figure 4.7). Then it attaches the hind end of its body with silk to the surface and suspends itself from the silk attachment, hanging vertically. It assumes a J-shape while hanging, and it stays this way for a day or more while its insides are turning into a pupa and a covering for the pupa. Check it periodically (see Figure 4.8). When you notice that the long black tentacles on the curved end of the J suddenly go limp and dangle freely, then you know

the caterpillar will become a chrysalis in the next fifteen to twenty minutes. (See Figure 4.9.) Stop everything and watch! To transform itself, the J-shaped caterpillar begins to wiggle violently and continues thrashing about for three to four minutes to sling off the larval skin. The skin then falls in a little wad underneath, revealing a lovely pale-green chrysalis dangling from same single silk attachment. It takes about twenty minutes for the chrysalis to assume its final shape, which is vaguely the shape of a pear. The wider portion is near the top. Metallic gold spots appear on the outer skin of the chrysalis, making it truly one of the most beautiful and ornate insect forms I've seen. The chrysalis hangs motionless for ten days to two weeks, depending on the temperature. (See Figure 4.10.)

The Monarch Butterfly Emerges

The day before the butterfly emerges, the chrysalis no longer looks pale green. Instead, the orange and black of the monarch butterfly can be seen through the translucent chrysalis shell. Usually during the early morning of the following day, the chrysalis shell splits and the butterfly slowly crawls out. The butterfly hangs upside down on the chrysalis shell with its legs, while its heart pumps blood into the vessels of the wings. Over a period of about two hours, the crumpled wings slowly straighten out and dry. When the drying process is complete, the wings will look mature, and the butterfly will begin moving around. You can offer it some sugar

Fig. 4.7 A monarch caterpillar preparing to form its chrysalis. It assumes a horizontal upside-down position, then its body condenses, getting shorter and more compact.

Fig. 4.8 A monarch larva in the characteristic J-shape before it sheds its skin to reveal the chrysalis underneath.

Fig. 4.9 A monarch larva in J-shape. Its tentacles have gone completely limp, signaling that metamorphosis into a chrysalis will happen within the next 20 minutes.

Fig. 4.10 A monarch chrysalis, pale green with a dazzling ring of metallic gold dots.

water or apple juice. Then, if it's not raining and not too windy, take it outside with the students. If you have nectar flowers, set it on one in full sun so it has an option to feed and can warm itself. When it's ready, it will take off. (See Figure 4.11.)

You can see videos of the entire pupation process on the Journey North website or on YouTube.

Observations of Caterpillars

Because monarchs are dwindling, I try to avoid handling them. When handled, they may appear to be unaffected, but then later they may fail to emerge from the chrysalis. On the other hand, bringing them indoors can be helpful. If you're attentive to their needs indoors, then bringing the eggs and caterpillars inside actually improves their chance of survival. I know this is true for the ones in my yard because they almost always disappear when left outdoors, presumably eaten by predators. (See Figure 4.12.)

Even without activities and experiments that involve handling them, you can still encourage students to observe them carefully. Below is a detailed description of a guided observation lesson for monarch caterpillars. My activist friend Beth Henry, who taught me much of what I know about monarchs, follows this sequence during her frequent forays into classrooms with her beloved monarchs in tow.

If you have monarch caterpillars in your classroom that students can observe, I suggest giving them a list of questions or prompts that force them to look carefully at the caterpillars—similar to what Beth does. Let students take turns looking at the caterpillar and jotting down answers to the questions or prompts, which you can discuss afterward as a class. You might give them a description of the different instars (from earlier in this chapter), and tell them they can look at videos on Journey North to help them answer questions.

For example:
- Describe the colors and patterns on a monarch caterpillar.

- Describe how one instar differs from another one.

- How are the two kinds of legs (the fleshy prolegs toward the rear, and the jointed true legs or thoracic legs behind the head) different?

- Do they use the legs differently in moving along a stem?

- Describe the motion of the head and maxillae (jaws) when the caterpillar is eating a leaf. (The head moves up and down, not side to side, as the jaws cut the leaf.)

- How is a monarch caterpillar different from the caterpillar of an eastern black swallowtail (or any other caterpillar)? The colors and patterns? Locomotion? Motion of the head and jaws when eating?

- How is a monarch chrysalis different from the chrysalis of an eastern black swallowtail (or any other chrysalis)?

Fig. 4.11 This recently emerged monarch takes nectar from milk-weed flowers.

Jennifer's Students, Inspired by Citizen Science

Jennifer Joyner is a teacher at Charlotte Country Day School who's gone the whole nine yards with monarchs in the classroom and has loved it! I talked with her about what she's done, how she got started, and what it's meant to her students. If you're interested in using monarchs at your school, you'll find her story useful and intriguing!

Jennifer teaches Spanish in middle school. She and the school's art teacher first got involved with monarchs

Fig. 4.12 Monarch caterpillars can easily be observed without handling them.

through the Symbolic Migration facilitated by Journey North (www.learner.org/jnorth/). In this program, American (or Canadian) students correspond with Mexican students living near monarchs' overwintering sites, by simulating a migration. At Jennifer's school, this starts each year when sixth grade art students create butterfly-sized paintings of monarchs as a study of symmetry in nature. Then Jennifer's sixth grade Spanish students write letters in Spanish to a group of Mexican students selected by Journey North. Jennifer mails the letters and paintings together (via Journey North) to the students in Mexico during the time of the real monarch migration to Mexico.

The Mexican students keep the butterfly paintings over the winter at their schools near the overwintering sites, and they mail butterfly paintings back to Jennifer

Fig. 4.13 Students' letters and their small paintings of monarchs are mailed to Journey North, to be forwarded to students in Mexico. Photo courtesy of Jennifer Joyner.

during the real spring migration, along with their own letters. In addition to this exchange, Jennifer's students read the Mexican students' blogs on the Journey North website. They learn about their counterparts' lives and the financial importance of monarch tourism for the families near the overwintering sites in Mexico. She says it has impacted the students' worldview and their understanding of similarities and differences in other parts of the world—quite an accomplishment for sixth graders. (See Figure 4.13.)

A couple of years ago, Jennifer decided to go beyond the Symbolic Migration and actually raise living monarchs. "I was looking for a way to expand our study of butterfly migration and make it more experiential and strengthen curricular connections." To take this next step, Jennifer attended a two-day workshop sponsored by the Monarch Teacher Network. She was accompanied by language arts/social studies teacher Debbie Biggers, who had already started a school garden based on her classes' study of the book *Seedfolks*. At the workshop, the teachers learned not only the specifics of growing milkweed and raising monarchs but also how students can tag the adult butterflies before releasing them, as part of scientists' efforts to track their migration and survival. Jennifer hoped this might help students develop a feeling of responsibility toward the natural world and a sense of empowerment from collaborating with professional conservation scientists—an unusual opportunity for youngsters. Or anyone!

Back from the workshop, the first step was to plant milkweed. With the help of a science teacher at their school, the two teachers who'd attended the workshop created two garden beds strictly for milkweed. From online vendors, they bought seeds and young plants of four milkweed species: swamp milkweed, common milkweed, butterfly weed, and tropical milkweed. (Rose Franklin's Perennials was one online source.) The teachers even dug up a few plants along busy roads. Their students learned how to plant and tend both the milkweed as well as nectar flowers in a nearby bed. The flowers and milkweed grew and, to everyone's joy, the longed-for monarchs came! The classes had their garden certified by Monarch Watch as an official Monarch Waystation. (See www.monarchwatch.org.) The plantings provide not only a nectaring rest stop for migrating butterflies but also an egg-laying site with lots of milkweed. As a bonus, their nectar garden has been named a Certified Butterfly Garden by the North American Butterfly Association (www.naba.org). Most adult butterfly species feed on nectar.

The flowers and milkweed have been thriving for a few years now. When a new school year starts each August, students are thrilled to find monarch butterflies on the nectar flowers, and tiny eggs and caterpillars on

Fig. 4.14 This monarch has been tagged with the round stick-on tag from Monarch Watch. Photo courtesy of Jennifer Joyner.

the milkweed! Jennifer says they are so eager to find new arrivals and to monitor the progress of the young monarchs that they check the garden before and after school, between classes, and during advisory and lunch!

At the Monarch Teacher Network workshop, Jennifer and Debbie had learned how to make caterpillar cages for the classroom by covering upside-down wire tomato cages (available at gardening stores) with a fine

Fig. 4.15 A monarch the class has raised lands on Olivia's face for a sip of juice before taking off. Photo courtesy of Jennifer Joyner.

mesh. They also made soft cages for newly emerged adults by covering two quilting hoops (available at craft

stores) with white mesh. When autumn comes, they place each caterpillar cage over a potted milkweed plant or some clipped sprigs of milkweed in water. (They don't cut sprigs of tropical milkweed, though, because that species wilts when cut.) When the students bring in leaves bearing eggs or some of the caterpillars from outdoors, the eggs and larvae are placed in one of the cages. This improves their survival rate, because predators often eat those left outside. The students can watch the growth process of the caterpillars in the classroom—each of the five stages (instars) of larval growth looks a little different. The students love to see the fully grown caterpillars "do the pupa dance" to make their chrysalises in the cages! About ten days later, when a butterfly emerges from a chrysalis, the students gather around to watch excitedly, marveling at "the miracle of life," as one student put it. After emersion from a chrysalis, an adult is transferred to the soft cage for a few days, where it has ample room to exercise its wings. Butterflies can be kept for close to a week if they are fed—the teachers and students feed them a nectar-like sugar meal of juice on a paper towel, which they suck through their proboscises.

Now it's time to tag. The tags and instructions are acquired in a kit from Monarch Watch (www.monarchwatch.org). Each tag is a small coded disk, sticky on one side, which adheres to the monarch's wing. The unique code for each tag applied to a butterfly must be reported to Monarch Watch, and if tagged butterflies are later found, then the code should again be reported, allowing scientists to track the butterflies' movements during migration. This helps identify places where nectar and milkweed are critically needed. To apply a tag, a student allows the butterfly to crawl onto his or her finger from the soft cage, rather than plucking it off the cloth and possibly damaging its legs or delicate wings. When the butterfly on a finger has its wings upright and touching, a student gently holds the wings together while another student carefully applies the tag to a very specific spot on the underside of one wing. (See Figure 4.14.) Then the butterfly is taken outside. Sometimes a teacher sprinkles juice on the hands and faces of students, allowing the butterfly to take a last few sips—which delights the students! When sated, the butterfly takes off. The students wave good-bye; some point south and call out, "That way to Mexico!" as the butterfly's powerful wings carry it away. (See Figure 4.15.)

During the first year that students raised monarchs to adulthood in autumn, they tagged and released fourteen adults for the fall migration to Mexico. The second year, they tagged and released about fifty adults. Jennifer said the first year she pondered whether to include tagging in the program. However, after opting to give it a try, she decided to continue it the second year because it meant so much to the students. "The kids buy into the idea of helping scientists preserve the migrations. They crave helping out. It gives them an opportunity to do good in the world and they embrace it." In addition to helping scientists, the students all know they are improving monarch survival rates by providing nectar flowers for the migrants and milkweed for egg laying, and by sheltering the eggs and larvae from predators. What positive learning experiences!

Most caterpillar care and tagging occurs during the first five weeks of school, in August and September, so those weeks are the most intense. Egg production slacks off when the weather begins to cool, and by late September, most of the larvae have matured, the young adults have emerged from the chrysalises, and they have been tagged. During the winter, the teachers still weave the monarchs into the curriculum. For example, in social studies, the monarchs are incorporated into a study of nomadic peoples who move from place to place in search of food. In language arts, the students write about a variety of experiences with the monarchs. The monarch project is easily incorporated into the sixth grade science study of communities and food webs, as well as potential units covering plants, migration, insects, and conservation topics. Habitat loss is the biggest threat to animals all over the planet, whether from climate change or human development. The loss of milkweed due to agricultural use of herbicides is a prime example of critical habitat loss threatening the survival of a species. Because the students know intimately and love monarchs, it's a perfect springboard into the biggest challenge of our time: human alteration and depletion of our fragile Earth.

Jennifer says the springtime arrival of the monarchs is not as busy as the fall season—there are fewer monarchs. But her Spanish class does a two-month unit on the monarchs in spring, during which they undertake a study entirely in Spanish of the monarch life cycle and migration, and then follow it with an examination of the lifestyles in rural central Mexico, where the monarchs overwinter. It's during this time that her students compare and contrast their lives with those of their Mexican peers by reading the Journey North Blogs. By the end of it all, her Spanish students are capable of extended discourse in Spanish on these topics. Jennifer hopes that this study helps foster a global awareness in her students and an enduring empathy for others who share our planet.

About Jennifer and Debbie and their school, I have to say: what a powerful project. I've worked in schools about twenty years altogether, and I've never seen teachers or classes as excited and enchanted by a living species, over a sustained period of time, as this group of monarch enthusiasts. I'm impressed, and I hope I have conveyed the potential for any teacher to duplicate what they have accomplished. Aside from benefits to the monarchs, this cross-curricular, hands-on conservation effort is planting impressions in those kids that will hopefully inspire them to care about other cultures, wildlife, and wild places for the rest of their lives. I hope so! At this time in our planet's existence, not much is more important than that.

The three sources that help Jennifer and her coworkers the most are the following:

Journey North facilitates the Symbolic Migration and the mailing of the painted butterflies and letters to the Mexican students to simulate migration. Journey North locates the Mexican recipients, so no effort is required on the school's end. As a teacher, you just read the instructions on Journey North for creating the symbolic butterflies and mail the finished products to Journey North to deliver for you. Journey North's website, as I mentioned earlier, is devoted to teachers and is loaded with detailed lesson plans about monarchs and other migrating animals. Among its other resources, the website includes the Mexican students' blogs that Jennifer's students read. Jennifer also uses the Journey North website in her classroom to track the live migration and read timely articles about the monarchs' progress as they migrate southward and then northward.

Monarch Teacher Network is a grant-funded program based in New Jersey. It offers the two-day monarch teacher workshops that Jennifer and Debbie attended to learn about growing milkweed and raising monarchs in the classroom. They also learned how to build caterpillar and butterfly cages and how to tag monarchs. Monarch Teacher Network provides workshop attendees a plethora of interdisciplinary teaching ideas, games, and support materials. In short, the workshop provides everything you need to create a living-monarch program—even milkweed seeds! Jennifer's school, Charlotte Country Day School, hosted the MTN workshop in June 2012 and hopes to do so again in the future. For a list of current workshops offered around the United States and Canada, check the MTN website.

Monarch Watch is the organization (and website) that actually creates and sells the tagging kits and keeps track of all reported data from the tags. The scientists at the University of Kansas who operate the site are also the ones who make use of the tag data to generate maps that can be seen on the Monarch Watch website. Plus, they use the tag data in conservation efforts. Monarch Watch also offers instructions for creating a Monarch Waystation and having your milkweed garden certified as an official Monarch Waystation, complete with a sign that identifies it as such.

From Beth Henry: Sharing Monarchs with Classrooms

Beth Henry is a naturalist, political activist, gardener, and butterfly enthusiast who's also my good friend. Beth introduced me to monarchs. She has planted and developed a mature native meadow and native forest (all native plants) on her suburban property. The meadow includes lots of milkweed and host plants for other butterflies, so her yard is full of monarchs and others.

Beth was featured in the book *Going Green* (Fulcrum Publishing 2008) that I cowrote with my daughter, Sadie, because of her volunteer work planting a native meadow and woodland at a local middle school. Beth has been a major force in my city in the fight to slow climate change. Taking monarchs to schools to educate youngsters on the butterfly's plight is one of her many services to our planet.

In Beth's own words, here is a description of what she does in the classes she visits:

Kids love butterflies, so doing monarch lessons with elementary classes is fun. I base the lesson around what I happen to have at the time: usually caterpillars, sometimes chrysalises too. I have a lot of milkweed and nectar flowers in my yard, and the plants attract the monarchs. I get a few monarchs during the spring migration northward and a lot more during the fall journey south, with almost none in June and July.

I originally bought the young milkweed plants as "deep plugs" from a mail-order native-plant nursery and put them in the ground in autumn. Being perennials, they need winter to get rooted and established before facing the intense summer heat of North Carolina. Some of my plants are common milkweed, which spreads aggressively, so it has now spread throughout the garden.

When I take the monarch caterpillars or chrysalises to school, I dig up some of the smaller milkweed plants and put them in pots to take with me. I like to show the students the caterpillars on the milkweed, so they can see their normal behavior, and so they can see what milkweed looks like. To survive the decline in their host plant, monarchs need for

everyone to plant milkweed, so teaching about milkweed is an important part of the lesson.

In addition to using living monarchs for the lesson, I use several posters that show the monarch life cycle, a map of the migratory routes, and pictures of the Mexican highlands where the monarchs overwinter.

To set up the lesson, I put the pots together on the floor. If I'm bringing chrysalises, they are usually attached to woody twigs or sticks, so I have the end of each stick taped to the side of a jar, or propped up in a pot of soil to keep it off the floor. I pull maybe ten chairs in a semicircle around the pots or jars, and a chair for me behind the pots. Then I have the students come in groups of ten to see the monarchs. If I have caterpillars, I ask the students to describe the caterpillars, which forces them to look carefully at the color pattern, at the two different kinds of legs, and at the black tentacles on the head and tail end of the older caterpillars. If I have different ages of caterpillars, I ask them to describe how the younger one differs from the older one. During this exchange, I tell them the names of any body parts they mention (true legs or thoracic legs, prolegs, tentacles). We talk about why the older one has a striped pattern, and we talk about camouflage. I ask the students what animals they think might want to eat or parasitize a caterpillar (parasitic wasps, praying mantises, birds). We observe a caterpillar eating, moving its head up and down as it clips the edges of

Fig. 4.16 A monarch chrysalis under the edge of countertop in Beth's house.

a leaf with its jaws. I ask the students to describe how the caterpillars walk: do they alternate right and left legs when walking, the way humans do?

After the students have observed the caterpillar closely, I tell them about the life cycle, using the poster: egg to caterpillar to chrysalis to adult and back to egg. If I have a chrysalis with me, I show it to them, and they usually marvel at how beautiful it is—pale green with metallic gold spots. It is

Fig. 4.17 Beth Henry and myself at the March on Wall Street South at the Democratic National Convention in Charlotte, North Carolina, in 2012. We were there with thousands of others to march for environmental justice and to speak out against the corporate drivers of global warming. Photo by Monica Embrey of Greenpeace.

pretty spectacular compared to the drab-colored chrysalises of most butterflies. I describe how, on the day before pupation, the fully grown caterpillar attaches itself by the tail end to a stick or other surface and then hangs vertically, gradually assuming a J-shape. About fifteen to twenty minutes before it sheds its larval skin to reveal the chrysalis, the black tentacles on the head suddenly go limp. That's a sign it's about to happen! Sometimes I leave the class with a larva that's close to pupation, so they'll have a chance to observe how the

caterpillar quickly wiggles out of the larval skin and slings it aside, revealing the delicate case of the chrysalis.

Moving on, I use the poster with migratory maps to explain monarchs' unique migration. I talk a little about the specific climate of the Mexican mountaintops where the monarchs overwinter, and how global warming is likely to shift the weather conditions there—possibly making it inhospitable to the monarchs. Then, to encourage the kids, I shift the focus to what they can do to help monarchs. Plant milkweed! I explain why milkweed is disappearing—mostly due to liberal spraying of the herbicide Roundup on cultivated fields in the Midwest. I tell them and their teacher about websites with instructions for planting milkweed and raising monarchs. Some of these sites, such as Monarch Watch and the Monarch Larva Monitoring Project, have citizen-science opportunities for anyone who wants to contribute data to monarch conservation scientists.

I finish up by describing my own backyard, a meadow I planted of 100 percent native plants, many of which are milkweed and nectar flowers for monarchs. I check the milkweed daily during the warm months. When I find monarch eggs or larvae, I bring them inside on their milkweed branches, to keep them safe from predators. I put the branches in jars or vases in the mud room/laundry room of my house, covering the opening of the jar so the larvae can't fall in the water and drown. I make sure they always have fresh milkweed. When the larvae are ready to pupate and form a chrysalis, they like to wander. I provide branched sticks, but they often ignore the sticks. Instead the caterpillars crawl all over the room, often choosing a spot on the underside of the cabinets or the ceiling! After they attach themselves and start the process, you can't move them without damaging them. So I watch and wait. After the butterflies emerge, unfurl and dry their orange-and-black wings, I release them outdoors near my nectar flowers. But I leave the empty chrysalis cases to decorate the room, because I love monarchs. The cases are happy reminders of the ones who've crossed my path, and the joy I feel in helping even one monarch make it safely to adulthood.

Bugs That Squeak and Hiss

Chapter 5
Bessbugs:
Glossy and Glamorous

Introduction

I love bessbugs! They may be my favorite insect in this book, other than perhaps praying mantises. They're not as stunning as the hissing roaches but are perhaps more charismatic—and much more docile. At 1.5 inches (3.8 cm) long, these beetles are big enough to be impressive... and they're beautiful! They're sometimes called patent-leather beetles because they're shiny jet-black like patent leather (see Figure 5.1.) I've never seen another insect so shiny. Other names for them include bess beetles or Betsy beetles—which makes sense, since they're classified as beetles.

Students are enthusiastic about bessbugs. They expect such a big insect to be intimidating and are thrilled to find the opposite. Bessbugs move very slowly with a comical lumbering gait, unable to move fast or fly (See Figure 5.1.). They also squeak when handled—an unusual and endearing trait in a beetle!

Odd Social Lives

Bessbugs have the ability to squeak because they live in "colonies," which is rare among insects. They squeak to communicate with each other. Most insects are solitary, other than the familiar social insects (Hymenoptera), which include bees, wasps, ants, and termites. Hymenopterans usually have different adult castes that perform different jobs in and out of the nest; castes may include workers, soldiers, a queen, and sometimes more. Each caste has its own anatomy and behavior.

Bessbugs don't have castes. To say that bessbugs are "colonial" just means that they live together in colonies, communicate with each other, and raise the young cooperatively. They create tunnels in rotting logs or stumps by chewing through dead wood, simultaneously getting their nourishment from bacteria and fungi in the wood. Adults live and raise the larvae inside the tunnels.

Fig.5.1 Lily admires the shiny, gentle bessbug

Fig.5.2 A bessbug. Note the striations on the wings. The sensory antennae are always active.

Females lay the eggs of course, but males feed the larvae. Bessbug larvae are plump white grubs—somewhat similar to caterpillars. I've never seen the larvae or pupae, because I don't tear their logs apart. I do know, however, that the larvae can squeak too by rubbing the tip of the third leg against a joint of the second leg. Adults produce the noise by rubbing a rough area under the wings against a rough area on the abdomen. All this squeaking is more properly called stridulation, as it's called in crickets.

No Danger to Wood Construction

Bessbugs eat only rotting wood. They are unable to eat wood that's not rotting and pose no danger whatsoever to human constructions. Even if a home had rotting wood in the walls or floors, bessbugs could not eat that— it's too dry, it's not on the forest floor, and it lacks the volume required for a colony.

Bessbug Conservation

These handsome and vocal beetles are native to the southeastern and mid-Atlantic states in the United States. They're not endangered but are not as common as they once were due to loss of habitat. When land is cleared to accommodate our growing human population, rotting logs and stumps are cleared away too.

To create habitat for bessbugs, logs must be left on the ground and allowed to rot. Not many homeowners are willing to do that, although more of us should. If you have an interest in getting your yard or schoolyard declared a Certified Wildlife Habitat by the National Wildlife Federation, leaving dead leaves and logs on the ground is an important step.

I used to find bessbugs easily in a small wooded area behind my house, but I know of only one colony there now. It may be because of the droughts we now commonly experience in summer, as our planet's climate changes. Bessbugs need moist wood. Or it may be due to "infill"—all of the undeveloped areas in my neighborhood have now been converted to homes and manicured lawns, due to population growth. Or maybe the increasing use of pesticides and lawn services has affected my neighborhood.

Even though my family doesn't use pesticides and we do leave rotting logs on the ground, animal populations can't survive in an isolated island—one unpolluted, unmanicured lot in a sea of development and toxic sprays.

How to Get Bessbugs

To locate bessbugs, you'll need to find some rotting logs on undisturbed land. Any time of year is as good as any other because they have a natural antifreeze that allows them to remain active in winter. Your search area doesn't have to be a big tract of land; a log in a neighborhood can work if it's in an area where the ground is covered with fallen leaves and logs as you'd see in a woodland. The logs and stumps where I find bessbugs behind my own home are in a wooded strip of land only about 50 feet (15.24 m) wide, between my yard and a ditch that borders a school playground. The ground cover there is leaf litter and sticks and invasive English ivy—no lawn, no plants that anyone tends to.

Teach Your Students to Respect the Needs of Wildlife

Be aware that it's illegal to remove any living thing from a government-owned park, whether it's a city park, a state park, or a national park or reserve, so please model for your students a respect for those laws. Parks and reserves are the only safe havens that wild things have left, as our human population grows. Poachers can and do sometimes decimate natural populations in parks, often for monetary gain. Very sad.

Homing in on Logs with Bessbug Colonies

In looking for bessbugs, you'll be rolling logs and looking around rotting stumps. You may find a couple of adults right under a log. I discovered them quite by accident that way, the first time I saw them. If none are immediately visible to you under the log, look for a clue to their presence: small piles of rolled up wood (feces) outside the log or under it. A single fecal pellet is brown, dry, round, and smaller than a BB.

If you do find a couple of beetles under a log or stump, put them in a container with some of the moist rotted wood. Two beetles are plenty for the experiments in this book, and the population in a given log or stump can recover from the temporary absence of two adults. In the classroom, students can work in groups and can take turns with the bessbugs. You'll need a couple of chunks of wood too, enough to roughly equal the size of a shoe. All in one piece is best, to allow for some tunneling. Find a second log that seems to be the same kind of wood as the bessbugs' log, then take a piece off of the second log instead of the beetles' residence. For carrying the beetles and wood back to school or to your study area, a clean bucket is fine; a plastic dishpan or shoebox will also work. Plastic is better than cardboard as it retains the moisture in the wood.

Be a Good Steward, Roll That Log Back

Whenever you roll over a log or a rock to see what's under it, please model good conservation practices to your students and put the log or stone back the way you found it. Even if the students aren't present, you can tell them what you did. As an ecology lesson, you can tell students that it may have taken years for the tiny plant, fungi, and animal communities that live under logs and stones to develop their habitat.

If there are any especially delicate creatures under the log such as a toad or a salamander, take care not to crush them. One way to do that is to pick up the animal and let it crawl back under by itself, after you've replaced the log—unless it's winter and the toad is hibernating! In that case, leave the toad where it is and put the log or stone slowly and carefully just as it was.

No Need to Tear Up the Log

If you find evidence of beetles and suspect they're in the log or stump but can't find them, it's okay to pull away a small piece of the log, leaving the piece intact so you can replace it. **The population can't recover if the log is demolished because they must have tunnels to care for their larvae.** And because unoccupied rotting logs may be far apart, chances are the beetles won't find another suitable log. As soon as you find a couple of beetles and get a piece of wood, there's no need to disassemble any of the logs further.

Materials You'll Need

Cages and food for captive bessbugs are described in the following section. In the latter part of this chapter, each experiment will list any extra materials you may need for that particular experiment.

How to House and Feed Bessbugs

Keeping bessbugs at home or school is easy. They need untreated wood from a rotting log or stump and they need moisture. I use oak; some sources say other hardwoods are okay, but I can't vouch for that. Don't use the wood of an evergreen such as pine—bessbugs won't eat it. The wood for the beetles must be in the state of

decay that wood is when it can be broken apart by hand. I try to give them an intact piece at least 4 to 5 inches (10.2–12.7 cm) in diameter and 8 inches (20.3 cm) long. That's not essential, but they do like to tunnel.

To set up a habitat, put 2 to 3 inches (5–7.6 cm) of damp (not soggy) soil in the bottom of a plastic or glass terrarium or a plastic dishpan. Any wood resting against the side must be at least 4 inches (10.2 cm) from the top of the pan so the bessbugs can't crawl out. Scatter broken-up leaf litter on top of the soil, then put in a piece or pieces of a damp (not soggy) rotting log—a piece that's big enough to have nooks and crannies for them to explore and big enough to crawl under. I give them a piece that's big enough to dig tunnels in, and they do. That might be 4 or more inches in diameter, maybe 8 inches (20.3 cm) long.

To keep the wood moist, I put on top of the log fragment a damp rag that has no cleaning-agent residue on it. I spray the rag every day with a plant mister filled with spring water from the grocery. I also feel the bottom of the wood every two or three days, and if it feels dry on the bottom, I drizzle a little water over the log. Remember to keep the wood damp, not soggy. If the log and soil dry out, the bessbugs will try (in vain) to climb up the walls of the terrarium or dishpan to escape. If you see that happening, they're in dire straits! Moisten the wood right away or they'll die of dehydration.

If the wood keeps drying out and you have trouble keeping it wet, or if the beetles eat up the original supply, add additional pieces of rotten branches a couple of inches in diameter from outdoors, still being mindful not to create a walkway for the beetles to the rim of the container. They seem appreciative of new wood from time to time.

Ordering Bessbugs

Yes, bessbugs can be ordered through the mail. I don't recommend this, however, because there's no guarantee where the beetles came from (a state park?). I know of no one who breeds the beetles in captivity, although it's possible. Carolina Biological says that their bessbugs come from their own extensive wooded "campus," but they acknowledge that they do accept various kinds of specimens from outside suppliers when needed. That makes me nervous. I worry about the future of bessbugs! If you do order bessbugs, ask to speak to the manager and find out where the beetles are coming from. Insist that you want the beetles to come from their own property, and that you are concerned about "overharvesting" of animals in general and about the decline in suitable habitats for bessbugs because of human developments. When they hear consumers voicing such concerns, vendors are much more likely to attend to these issues.

Field Hunt

I don't generally take students on field hunts for bessbugs because I don't want to give them ideas about looking for the beetles on their own later, and tearing up logs. I feel too protective of the beetles and other creatures that live in and under our diminishing supply of rotting logs in natural areas.

Bessbugs at School

Observations and Activities

Place the terrarium in your classroom for a few days before doing any experiments, and ask the students to jot down observations they make. It's safe to leave the lid off the terrarium as long as the wood doesn't provide a path to the rim and as long as you keep it moist by spraying the rag and drizzling water on the wood as needed. Bessbugs definitely can't climb a vertical expanse of plastic or glass.

As the students try to make observations, the bessbugs may stay totally out of sight in or under the log fragment(s) in the terrarium. If they do, you could gently pull the beetles out and put them on top of the log, then allow the students to watch what they do. (The beetles will immediately try to head back under the wood.) That gives you a lead to ask the students: What is it about the top of the log they don't like? Someone may suggest they don't like light or the drying effects of air—or that they prefer to be hidden. A student may propose they just have an instinct to burrow in wood, and given a chance, that's what they'll do. Young students are more likely to say that the beetles are scared, or they want to go home, or something along those lines, which is fine.

Thigmotaxis and Phototaxis

Bessbugs give the impression that they have an aversion to light (negative phototaxis) and that they prefer cover and close surroundings (positive thigmotaxis). For example, if you put a bessbug on top of a jumble of damp rotting wood in a terrarium, the beetle will immediately begin trying to get into and under the wood. But is this really phototaxis or thigmotaxis? Maybe not! This will be explored in the Experiments section.

Intriguing Antennae!

When exposed to new air, a bessbug waves its antennae around, as if trying to detect scents in the air, or perhaps air currents. They do this when removed from their terrarium and put on a new surface, when taken outside, when exposed to air from an open door or window. Make sure students notice that the waving of the antennae is not constant, but only when the air around the bugs is new. The antennae aren't long—length could be a problem in the tunnels where they live. But the comblike structure increases the surface area for sensing chemicals. (See antennae in Figure 5.6.)

Sometimes bessbugs also tap the ground in front of them with the antennae. For example, in the T-maze experiment that follows, a bessbug will often tap the ground with the antennae when it reaches the end of the long stem of the T and must decide whether to turn left or right.

I'm not sure I've seen bessbugs tap each other's antennae when they meet, the way ants do. Ants definitely communicate by touching antennae, perhaps just identifying one another as nest mates. Ants also use their antennae to pick up scent trails left by their nest mates—trails that lead to food sources. My impression is that bessbugs use the antennae more to gather information about changes in their environment than for communication.

Bugs Shocked by Human Breath!

I was recently discussing the function of bessbug antennae with a couple of teachers, when something interesting happened. In the midst of the conversation, one teacher breathed very slowly (with an open mouth) on one of the bessbugs, just inches away from it. We wanted to watch the bug wave its antennae around in response. Instead we were stunned to see a violent reaction! The bessbug immediately recoiled in every way possible: it jerked its body back, it stepped backward as fast as it could go, it lay its antennae flat, and it squeaked repeatedly! What the heck? I've never seen a bessbug react in such a way to anything! (This person had normal breath, mind you!) Upon reflection, I'm sure the beetle did detect either warmth or scents in the breath by way of its antennae, even though the antennae were quickly drawn back in reaction. As for the violence of the reaction, I'm guessing that bessbugs have evolved to avoid the breath of a mammal—a potential predator (such as a bear or raccoon ripping into their home log). Any mammal tearing up a log looking for grubs or other insects would probably be sniffing vigorously, possibly snorting and huffing, to find the insects, and thus its breath would be detectable, by scent or warmth or both. Given that, the recoiling is understandable.

But what about the repeated squeaking? Wouldn't the noise alert the predator to its presence? I see only two possibilities for why it would be adaptive to squeak in the face of predator. One is the possibility that a predator would be confused for a moment by the squeaking (few insects squeak), thus giving the bessbug time to escape. The other possibility is that squeaking could alert others in the bessbug colony of a predator, thus giving them time to sequester the vulnerable larvae or retreat to the deeper tunnels in the log. Many animals are known to give warning signals to others that may include family. Rabbits and deer flash the white area on their tails when fleeing danger. Prairie dogs and meerkats vocalize to colony-mates when a potential predator is spotted. Many bird species give warning calls when a threat is near.

Curiously, if you blow on a bessbug forcefully, it doesn't elicit the same response. I suppose predators are likely to exhale when ripping into a log, but unlikely to forcefully blow.

Another Way to Produce a Squeak

If a student wants to hear a squeak, you can hold a beetle close to his or her ear and jostle the bug a little by tapping its back very gently with one finger. Usually that will produce a squeak or two. Is it a warning squeak to colony-mates, or a squeak to startle a potential predator? Difficult to say.

Do a Lesson on Overharvesting and Poaching

Bessbugs are a good animal to use for a conservation lesson about threats to wildlife in a world where climate is changing, habitat is declining, and human development is growing rapidly. Did you know illegal trafficking in wildlife and wildlife parts is the third biggest black market in the world, behind drugs and weapons? Poaching, or the illegal harvesting of protected animals, is huge—for the pet trade, for food, souvenirs, clothes, for medical research, and for the superstitious use of animal parts in traditional medicines, especially in China and other parts of Asia but also in Africa and the Americas—even in the United States. For more about that, see the Preface to the Second Edition. See also the website TRAFFIC: The Wildlife Trade Monitoring Network (worldwildlife.org/initiatives/traffic-the-wildlife-trade-monitoring-network). As I mentioned earlier, collection of any plants or animals from parks and reserves is illegal. Overharvesting of a declining species from anywhere is bad news. Modeling good stewardship to your students can include returning your bessbugs to the same log you got them from, after you've had time to investigate them.

If returning them to their home log is not possible, try to find another rotting log near that site. But if you don't live near their place of origin, it's best not to release them at all. It's also not a good idea to release them in the grass somewhere. Other options include finding another teacher who wants them; bessbugs are ideal for teaching children (and squeamish adults) about insects. Donate them to a local nature museum. Or consider keeping them indefinitely as unusual classroom pets!

Education is important in protecting our planet and the animals who share it with us. Ask your students to tell their friends and families about these intriguing beetles, the problems they face, and how people can help. Awareness precedes action! People won't get excited about saving animals if they don't know anything about them.

Students Can Nurture Wild Bessbugs

What can you and students do, at home, to help bessbugs and their fellow denizens of rotting wood?

In your own yards (or in the schoolyard), leave dead hardwoods and stumps standing. Leave logs and thick branches of oak or other hardwoods in moist, shady spots on the ground, in places not affected by pesticides. Logs 2 to 3 feet (.6–.9 m) long are long enough, but longer is fine. Over time, the logs will begin to rot, and small creatures will utilize the logs. Hopefully, some will be bessbugs! If not, no doubt something interesting will come! You'll also be providing food for the many bird, mammal, reptile, and amphibian species that eat insects.

Experiments for Bessbugs

Some of these experiments are designed to explore the bessbug's niche or its relationship with its native environment. These experiments highlight the characteristics of the bessbug that help it survive in the wild. And some of the experiments are designed to provide a fun way to test hypotheses.

Remember that the hypotheses I give are just examples. Your hypotheses will be the predictions made by the class or a particular student. Your result for each experiment will be a statement of how your animals reacted to your experimental setup. Your conclusion is a statement of whether your prediction was confirmed or not. For each experiment, adding replicates increases your confidence in the validity of your conclusion, but they may be omitted if tedious for young children.

Experiment 1

Students love racetracks; racing appeals to their competitive nature. How often do students race among themselves to see who can be first in line, who can finish first in sports competitions? Whether racing each other or racing bugs they've taken charge of, kids love to root for their predicted winner!

Question:
How fast can a bessbug move on the racetrack? Can a bessbug beat a hissing roach? (Consider reading the fable "The Tortoise and the Hare" to your students before trying this competition!)

Hypothesis:
Each student can predict how much time the bessbug will take to move from start to finish on the "racetrack." For example, "If we place our bessbug in the center of our racetrack and start timing when the bessbug starts walking, it will cross the finish line in an average of twenty seconds." Or if pitting a bessbug against a hissing roach, each student predicts which one will win. If multiple trials are planned, students predict how many trials each insect will win.

Materials and Methods:
See Experiment 1 in Chapter 6 on Madagascar hissing roaches for directions on how to make a simple racetrack on paper. Exactly the same racetrack can be used for this experiment. The procedure for conducting a solitary race is also described in Experiment 1 for hissing roaches, and the same procedure can be used for bessbugs.

To time a bessbug in a solitary race:
I've had students work in groups of four students, with one racetrack per group. One student in the group can operate the stopwatch or watch the clock, one can record results, one can release the bessbug when everyone's ready, one can retrieve the bessbug at the end. If you have only one bessbug, this can be

done as a class activity, or student groups can do the racetrack experiment in sequence.

When it's time to race, trap the bessbug under a clear plastic cup placed inside the inner circle of the racetrack, not overlapping the line. After the cup is lifted to begin the experiment, the bessbug may wander around aimlessly for a minute or so, "sniffing" the air with its antennae. But it will soon begin to walk in a straight line. When this happens, you can move the bessbug back inside the inner circle, and it will probably continue walking in the same straight line. You can start the timer when it crosses the perimeter of the inner circle. Or you can start the timer when you lift the cup initially, and include any wandering as part of the time to completion.

The results are more meaningful if you can do a number of replicates or trials (have the bessbug run the race several times). Doing ten repetitions is optimal, but even three is better than one. As part of scientific methodology, decide before you start how many trials you're going to do.

Fig. 5.3 The bessbug and hissing roach are set to race. Who will win?

Students can use the worksheet on page 68, which I designed for use in a solitary race or a paired race.

To pit bessbug against roach in a competitive race:
Bessbugs and roaches are fun to use as competitors because they're both big insects. This pairing is a student favorite. To set up the race, I put a transparent plastic cup over the roach and bessbug together inside the inner circle. The two are totally indifferent to one another when trapped together under the cup. If they start simultaneously when the cup is lifted, the roach will finish first every time.

An interesting variation on this pairing is to give the bessbug a chance to win by starting the clock at the moment the cup is lifted. If the roach doesn't start moving right away (which sometimes happens), the bessbug may win—which brings to mind "The Tortoise and the Hare" fable! The tortoise plodded along slowly but steadily, while the rabbit goofed off and lost. Or they may both take their time to get moving in a straight line toward the outer circle, in which case, the outcome is a toss-up.

You can do this as a class activity or have students work in groups if you have enough bugs. If you have only one of each bug, they can still work in groups, doing the experiment sequentially. If you use the data sheet on page 68, decide if you want each student to fill one out or each group to fill one out. Or you can fill it out as a class. (Or skip it!)

I've done this with groups of four, with duties assigned as follows: Student 1 releases the bugs

Fig. 5.4 Sophie expresses surprise as the bessbug beats the hissing roach, while Louie manages the timer.

for each trial by raising the cup and saying "Go!" or "Begin!" Student 2 records the winner for each trial on his or her data sheet (p. 68). Student 3 captures the bessbug at the end of the race and places it back under the cup for the next trial. Student 4 captures the roach at the end of the race and places it back under the cup for the next trial. After all trials are completed, Student 2 shares his or her tallies on the data sheet with the other members of the group so they can record the numbers on their own data sheets—unless they're using one data sheet per group, in which case they decide on a group hypothesis at the beginning and jointly compose the statements for the result and conclusion after the race.

Again, results are more meaningful if students conduct a number of trials. Ten repetitions is optimal, but even three is better than one. As part of scientific methodology, decide before you start how many trials you're going to do.

Result:
Students' results are a statement of their own observations, recorded times, or tallies.

For the solitary bessbug run:
If students record the time to completion for ten solitary runs from start to finish, then they should average the ten time measurements and compare this average time to their predicted time to completion.

In my experience, bessbugs on a laminated track cover the distance from start to finish in six to seven seconds. They are slightly faster on an unlaminated paper track, completing trials in five to six seconds. These times do not include time spent standing still or time spent wandering around the center while testing the air with the antennae.

For a competition:
If a bessbug and roach are competing, then the results are a statement of how many trials each individual "won," and which competitor won the most trials.

As mentioned in Chapter 6, a hissing roach can easily complete the race in one second. In my experience, bessbugs require five to seven seconds to complete the race if they begin walking right

Student name_____

Animal Racetrack

Question: How fast is a _____? Or, which animal is faster, a _____ or a _____?

Hypothesis: I predict that the _____ will cross the finish line in an average of _____ seconds/minutes.

OR

Hypothesis: I predict that _____ will win the race in _____ trials out of a total of _____ trials.

   Or, I predict that the _____ and the_____ will tie in the number of trials won.

Methods: We put _____ _____ in the center of the inner circle. We released it/them.

For a single insect: We clocked the time taken to reach the finish line (the outer circle). We repeated the procedure _____ times.

For a race between two insects: We made a tally mark in the appropriate space for each competitive trial.

Results:

_____ (name of Bug 1)   _____(name of Bug 2)

Put a tally mark here when Bug 1 wins a trial   Put a tally mark here when Bug 2 wins a trial

Total tally marks for this bug _____   Total tally marks for this bug _____

Class total _____   Class total _____

Conclusion:

Fig. 5.5 Student Data Sheet.

—— 68 ——

away. So if they both start moving right away, a roach will almost always win. The caveat is that the roach may space out under the retaining cup, sit still, and let the bessbug win! But… bessbugs may stall too before starting toward the finish line.

Conclusion:

The students accept or reject their own observations here, and state any conclusions they may have about the insects' behavior. Bessbugs don't need to be fast. Since they live in sequestered tunnels in logs, I imagine they are seldom pursued by predators. Lumbering around works fine in their particular niche. However, hissing roaches live on the forest floor in Madagascar. One reason for roaches' amazing success as a family is their speed, which helps them escape would-be predators.

Experiment 2

Question:

Who will win in a racetrack contest between a bessbug and (1) a mealworm, (2) an adult mealworm beetle, (3) a pill bug, (4) a termite, (5) an ant, or (6) a swallowtail caterpillar?

Hypothesis:

For each pairing, every student makes a prediction about which bug will cross the finish line first. If you want them to, they can also predict the time each bug will take from start to finish.

Materials and Methods:

See Experiment 1 in Chapter 6 on hissing roaches for directions on how to make the racetrack. Exactly the same racetrack can be used for any pairings of bugs in this experiment.

After you've decided which insect will challenge the bessbug in a race, students should predict how many trials each bug will win. Or they could predict how long each competitor will take, depending on how you plan to do the experiment. For any pairing, ten repetitions is optimal, but modify that number as needed.

Each student or group can record a hypothesis, and after the trials, can record the results and conclusion on the data sheet on page 68 (the same sheet used for Experiment 1 in this chapter).

To set up a race between a bessbug and its competitor, the two bugs can be placed together under a plastic cup inside the inner circle of the racetrack. All of these bugs are indifferent to one another when trapped together under the cup—no need to worry about "fights."

This can be done as a class activity, or students can work in groups of four. If you have only a couple of bugs, groups can do the experiment sequentially.

If you want the students to simply tally the winner for each trial, duties in a group of four can be assigned as follows: Student 1 releases the bugs for each trial by raising the cup and saying "Go!" or "Begin!" Student 2 records the winner for each trial on his or her data sheet (page 68). Student 3 captures the bessbug at the end of the race and places it back under the cup for the next trial. Student 4 captures the challenger at the end of the race and places it back under the cup for the next trial.

After all trials are completed, Student 2 shares his or her tallies on the data sheet with the other members of the group so they can record the numbers on their own data sheets—unless they're using one data sheet per group, in which case they decide on a group hypothesis at the beginning and jointly compose the statements for the result and conclusion after the race.

If you want the students to record each competitor's time for each trial run (rather than just recording the winner for each trial), you might assign duties a little differently for a group of four. Student 1 releases the bugs for each trial by raising the cup and saying "Go!" or "Begin!" Student 2 records the time to completion for the bessbug. Student 3 records the time to completion for the challenger. Student 4 captures both insects at the end of the race and places them both back under the cup for the next trial. After all trials are completed, recorded times are shared among the group, if individual data sheets are being used.

Result:

Students state their observed outcome, reporting the number of trials in which each competitor "won." Or if the competitors were timed, students report the average time for each competitor, declaring a "winner."

On my racetracks that have 7 inches (17.8 cm) between the starting line and the finish line (the perimeters of the inner and outer circles), I've observed these times:

Fig. 5.6 A bessbug—you can see the eyes on the edge of the head, the closed mandibles for chewing wood, the comb-like antennae, and front legs modified for moving chewed-up wood.

Bessbug: six to seven seconds on a laminated track, five to six seconds on a paper track

Hissing roach: one to two seconds, laminated or paper track

Ant: eight to ten seconds on a laminated track. Ants are the only insects I've tested that have never dawdled on the track, but always get going right away.

Roly-poly (pill bug): fifteen to seventeen seconds on a laminated track after it starts moving. A roly-poly usually sits still for about five seconds when placed on the track.

Adult mealworm beetle: four to five seconds on a paper track if it goes in a straight line

Adult mealworm beetle: ten to eleven seconds on a laminated track if it can figure out how to walk on the slick surface. Figuring this out can take a couple of minutes, and some never do figure it out.

Mealworm: thirty-two seconds—170 seconds on a laminated track, depending on whether it moves in a straight line

Termite: can't get any traction at all on a laminated track. On a paper track, I've clocked termites from 22 to 198 seconds, depending on whether they walk in a straight line (generally not).

Swallowtail caterpillar: five minutes on a laminated track after three minutes of sitting still in the center of the racetrack. On a paper track, 137 seconds with no stalling. Your students' results could be different, of course.

Conclusion:

Students write any conclusions they draw from the experiment. It's interesting to me that some insects have no trouble walking on a laminated surface and some do. The termites and some of the adult mealworm beetles were completely unable to get any traction at all. Different species have different kinds of legs adapted to their natural substrate.

Fig. 5.7 A roly-poly is a good competitor on the racetrack.

Experiment 3

Question:

Do bessbugs have a preference in a T-maze? Does the width of the T-maze affect their choice?

Hypothesis:

We think bessbugs will turn right and left equally in multiple runs through a T-maze.

Materials for two bessbug T-mazes:

1. Two pieces of stiff cardboard, each 7 inches by 8 inches (17.8 x 20.3 cm).

2. Six intact craft sticks from a craft store (a craft stick is the exact same thing as a wooden Popsicle stick). Each uncut craft stick should be $4^1/_2$ inches (11.4 cm) long and $^3/_8$ inch (1 cm) wide. These are referred to as "A strips" in the following construction instructions.

3. Two pieces of a craft stick, each 1.5 inch (3.8 cm) long and $^3/_8$ inch wide (referred to below as "B strips").

4. Two pieces of a craft stick, each $1^5/_{16}$ inch (2.4 cm) long and $^3/_8$ inch wide (referred to below as "C strips").

5. One hot-glue gun.

Method:

On each sheet of cardboard, you'll be constructing a T-shaped corridor using craft sticks. You'll glue the sticks on edge to create the corridors. Before you glue anything, draw the appropriate T-shaped corridor on each piece of cardboard, because it's difficult or impossible to pull hot glue off of cardboard without ripping it.

To give you the overall picture before you start gluing: you'll be starting the bessbug at the bottom of the T. The bessbug will walk up the long stem of the T, and then will turn either right or left at the top to exit through one side of the T. Students will predict and record which way it turns, over multiple trials. This procedure will be repeated with a slightly different T-maze, one in which the long stem of the T is a narrower corridor. As you'll see, the width of the long corridor affects the results, for reasons I'll explain. I'll describe T-maze no. 1 first. The long corridor on T-maze no. 1 is wider than on T-maze no. 2.

On T-maze no. 1, the bottom of the T begins ¾ inch from the lower end of the cardboard. Glue two uncut craft sticks ("A strips") parallel to each other, creating a long corridor 1¼ inches (3.2 cm) wide. The long corridor is parallel to the long axis of the sheet of cardboard. At the top of the two long sticks, glue two pieces of craft stick ("B strips"), intersecting the long sticks at right angles and extending out sideways to make the lower part of the T's crossbar. The corridor at the top is 1¼ inches wide, so glue an uncut craft stick ("A strip") near the top of the cardboard sheet, parallel to the "B strips" and creating that top corridor 1¼ inch wide. I write the letters L and R on the cardboard at the respective exits from the T-maze just to be sure students don't get left and right mixed up.

On T-maze no. 2, the bottom of the T begins ¾ inch from the lower end of the cardboard, as for the first T-maze. Two uncut craft sticks ("A strips") are glued parallel to each other, creating a long corridor only $^5/_8$ inch (1.6 cm) wide, much narrower than in the other T-maze.

Again, the long corridor is parallel to the long axis of the sheet of cardboard. At the top of the two long sticks, I've glued two pieces of craft stick ("C strips"), intersecting the long sticks at right angles and extending out sideways to make the lower part of the T's crossbar. The corridor at the top is 1¼ inches wide, so an uncut craft stick ("A strip") is glued near the top of the cardboard sheet, parallel to the "B strips" and 1¼ inches above them. I again write the letters L and R on the cardboard at the respective exits from the T-maze.

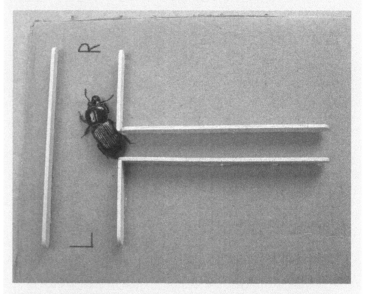

Fig. 5.8 Bessbug in T-maze no. 2, which has a narrow entry corridor that eliminates any opportunity for wall-hugging.

I do this experiment in groups of four so I don't have to make T-mazes for every individual. If you're industrious enough to make two mazes for every individual, at least they last a long time. If the craft sticks pop off, you can easily glue them back on. I've had my mazes for years. Even if a corner of the cardboard gets bent, they're still good.

Alternatively, you can do a class experiment with just two mazes. Or you can let student groups work in sequence, each group doing a few trials with each maze. Each student should make his or her own prediction of the outcome.

To begin, someone places the bessbug at the base of the T, gently guiding it to enter the T. Then students watch to see which way the bessbug turns at the top. Do ten trials with each maze.

Result:

Student results are a statement of the number of left turns and the number of right turns with each maze.

I first did this experiment with T-maze no. 1—the one in which the long corridor is wider. I noticed after a while that the bessbug tended to walk along either the right side or the left side of the corridor, in contact with either the left wall or the right wall. If it walked down the left side of the long corridor, then it turned left at the top. If it walked down the right side of the corridor, then it turned right at the top. Which way it turned at the top seemed to be completely dependent on which wall it happened to make contact with first in the long corridor.

It's interesting to see if students notice that wall contact upon entering the maze determines which way the bessbug turns, and if they notice that the bessbug seems to want to hug the wall. It's my impression that they do it every time. This makes sense, because in nature they live in narrow tunnels.

Hissing roaches, in contrast, don't do this. A hissing roach will use its antennae to maintain contact with both walls while walking down the long corridor of a T-maze, but they don't hug either wall, and contact with the wall has no effect on which way they turn. (Roaches require a larger maze, described in Experiment 7 of Chapter 6 on hissing roaches.) Students may notice the bess-

bug's behavior if you give them hints or if they've observed another insect in the T-maze that doesn't act this way (such as a hissing roach, mealworm beetle, caterpillar, or roly-poly). One hint you might give them is to suggest that they introduce the beetle into the long corridor on either the left or right side of the corridor, initiating contact with one of the walls at the outset.

Fig. 5.10 Sophie's bessbug turns left, which is happily what she predicted!

After noting the wall-hugging behavior, I made T-maze no. 2. In this maze, the long corridor is just wide enough to accommodate the bessbug's body. So both walls hug the bessbug as it proceeds up the long corridor. In this maze, the direction turned is not influenced by the contact with either wall, and the bessbug's choice seems random. No preference is noted.

Conclusion:

Students note any conclusions they draw from observing the bessbugs (and perhaps other insects) in the T-mazes.

Fig. 5.9 Louie watches the bessbug turn left in the T-maze.

Fig. 5.11 Sophie and Louie watch as the bessbug pauses before turning. Which way will it go?

Experiment 4

Question:

In an empty terrarium, will a bessbug exhibit wall-seeking behavior? That is, will it choose to position itself next to the wall most of the time?

Hypothesis:

We think the bessbug will hug the wall, or stay next to the wall most of the time.

Materials:

1. One square or rectangular box or terrarium per group. The box should be big enough that the area of the floor is 144 square inches or larger (929 sq. cm). This could mean a box 12 inches by 12 inches (30.5 x 30.5 cm), 10 inches by 14.5 inches (25.4 cm x 36.8 cm), or 8 inches x 18 inches (20.3 cm x 45.7), for example. The walls and floor of the box should be smooth with no folded flaps of cardboard. Otherwise, the beetle will be distracted with exploring a potential opening.

2. A piece of blank paper bigger than the floor of the box or terrarium, one for each box or terrarium.

3. A black marker.

4. A ruler.

5. A clock or timer for each group.

Methods:

Cut the piece of paper to fit the floor of the box or terrarium exactly. With a black marker, draw a rectangle or square on the paper that is 2 inches (5 cm) from the outer edge of the paper at all points to create a 2-inch-wide pathway along the outer edge of the paper. Put the paper back into the box or terrarium. If you're doing this as a class experiment, you'll need to use a terrarium with transparent walls or a box with low walls so that everyone can see its floor. Walls that are 2.5 inches (6.3 cm) tall are adequate to keep bessbugs from climbing out. If you have a large group and several bessbugs, students can work in groups. Each group will need one bessbug, one box with a pathway inside, one clock-watcher, and at least one person to record data. To begin, the bessbug is placed in the box. Give the beetle five minutes to recover its composure from being moved and to situate itself. After five minutes, either you or the clock-watcher in each group begins watching the clock and calling out thirty-second intervals. Students not watching the clock can record the beetle's position (between the black line and the wall, or not between the black line and the wall) every thirty seconds. Ten minutes of observations is enough—you'll have twenty observations per beetle at that time.

Result:

In writing up the results, students state the number of observations "next to the wall" and "away from the wall." During this time, the students should also jot down anything they notice about the beetles' behavior. They can also write any observations they may have, such as "the beetle sometimes stood on its back legs and tried to climb the wall." In my experience, bessbugs will stay between the black line and wall about 90 percent of the time in the box or terrarium. I've watched them for at least an hour and this behavior continues.

Conclusion:

Students either accept or reject their own hypotheses, then write whatever conclusions they may draw from the data collected. They try to make sense of whatever they observe and try to relate the observations to the beetles' way of life. Bessbugs live in tunnels in logs, so in nature, they are constantly in contact with walls and are probably instinctively drawn to that. But lots of animals will stay next to a wall in a strange situation. The wall provides cover or protection from at least one side. Some students may conclude that the beetles are next to wall because they're trying to get out, and that may be right too. I read a study about wall-seeking in humans, and you can see this for yourself. If a restaurant has tables distributed evenly throughout a large room, the tables along the walls tend to be occupied first. Is it instinct, to protect ourselves or avoid being conspicuous? Who knows?

Fig. 5.12 This bessbug hugs the wall as it walks around the perimeter of the box.

Experiment 5

Question:

Are bessbugs phototaxic—that is, attracted to light or repelled by light? Specifically, can we manipulate which way a bessbug turns in a T-maze by putting a light at one of the exit corridors?

Hypothesis:

Since bessbugs live in dark conditions, I predict that they are repelled by light.

Materials:

1. T-maze no. 2 from Experiment 3, above (instructions for making it are in Experiment 3).

2. One small working flashlight. I use a flashlight only 4 inches long, powered with three AAA batteries. I cover the lens with a piece of cloth secured with a rubber band to dim the light. The resulting light is dim enough that I can look at it directly with no discomfort at all. (Take care that students never look directly at an LED flashlight.)

3. A flashlight similar to the other one in use. But this second one won't be turned on, so it doesn't need batteries.

Method:

Put T-maze no. 2 on a flat surface. Put the lit flashlight about an inch away from one exit of the T-maze. Put the unlit flashlight in the same position on the other side of the T-maze. The exits of the T-maze are the two openings at either end of the crossbar at the top of the T.

I do this experiment in groups of four, because I don't have T-mazes for every individual. Alternatively, you can do a class experiment with just one maze. Or you can let student groups work in sequence, each group doing a few trials. Each student should make their own prediction of the outcome.

To begin, someone places the bessbug at the bottom of the T, gently guiding it to enter the long corridor of the T. Then students watch to see which way the bessbug turns at the top. Do ten trials if doing it as a class; if students are doing it in simultaneous groups, each group can do ten trials. But if you're using the same bessbug for sequential groups, do only three or so trials per group, and use the class total to assess the results.

It's a good idea to exchange the flashlight positions after every couple of trials to rule out the possibility that the bug is just choosing a consistent direction for other reasons.

Result:

Student results are a statement of the number of left turns and the number of right turns, and any other observations they may have made. When I have done this experiment, the bessbugs I've used have turned toward the flashlight about 85 percent of the time overall, which has surprised me. I've done it in dark rooms and in fully lit rooms. It's possible your students will get different results.

Conclusion:

Students draw their own conclusions about their results. I can't explain my own results. I expected bessbugs to be repelled by light. I would not have been surprised if they had been indifferent to the flashlights. But I was surprised that they seem to be attracted to light. How can this serve them?

The only explanation I can think of is this: when in the maze, the beetles are in distress because of the dry conditions there. I know that bessbugs will seek to leave their terrarium if the terrarium gets too dry. They try to leave by exiting the wood and trying to climb the walls of the terrarium (to no avail)—I've seen this. I know too that if I have bessbugs in an empty dry box, and I put a balled-up cloth in the box, they will climb on top of the cloth as if trying to get a boost out of the box. They don't seek cover by going under it.

So perhaps it is adaptive for them to seek a way out when conditions are too dry. If you live in a tunnel in a log, and the log is getting too dry, then the best way to find the exit is to walk toward the light.

Experiment 6

Question:

Are two bessbugs in an empty container attracted to each other, repelled by each other, or indifferent to each other?

Hypothesis:

Since bessbugs are colonial, I think two bessbugs will hang out together in an empty box.

Fig. 5.13 Louie carefully transports the bessbug to the experimental set-up.

Materials:

1. An empty terrarium or a rectangular box at least 12 by 18 inches (30.5 x 45.7 cm) or so. Find a box whose floor dimensions are multiples of two if possible, such as 12 by 18, 14 by 20, or 10 by 20.

2. A piece of paper at least as big as the floor of the box to cover it.

3. A black marker to draw a grid on the piece of paper.

Methods:

To test the hypothesis, you'll be putting two beetles on a grid of squares and tallying every thirty seconds whether they're in the same grid square. If they're attracted to each other, they'll be in the same grid square more often than you would predict if their location was totally independent of one another.

If you're doing this as a class experiment, you'll need to use a terrarium with transparent walls or a box with low walls so that everyone can see its floor. Walls that are 2.5 inches (6.3 cm) tall are adequate to keep bessbugs from climbing out. After you have a box or terrarium, you need a grid of squares on a piece of blank paper, which you can easily draw. Start by measuring the floor of the box and then, with a pencil, draw a rectangle on the paper the exact size of the floor of the box. You'll be cutting along that pencil line after you make the grid, but it's easier to draw the grid before you cut out the rectangle.

If the dimensions of the floor of your box are multiples of two, then start creating the grid by drawing a series of parallel lines on the piece of paper that are 2 inches (5 cm) apart. Use a black marker. Then rotate the paper 90 degrees and do that again. The second series of parallel lines should be perpendicular to the first, creating a grid of squares that covers the entire piece of paper. Each square will be 2 inches along each of its sides.

If your box dimensions are not multiples of two, then adjust the distance between the lines a little so that lines are still equidistant and are a little more or a little less than two inches apart. It doesn't matter if the resulting squares are not exact squares, as long as they're all the same size.

After you've drawn the grid, put it in the box or terrarium so that it covers the floor.

If you have several bessbugs, students can work in groups. Each group will need two bessbugs, one box with a grid on the floor, one clock-watcher, and at least one person to record data.

Fig. 5.14 This bessbug has a friend in Lily.

To begin, you or students put two bessbugs in the box next to each other and wait five minutes to give the beetles time to situate themselves. During this time students observe the beetles' behavior. Tell the students to notice how the beetles move their antennae, any apparent interactions between the two bessbugs, and any other movements they notice.

After the five-minute wait, either you or the clock-watcher in each group begins watching the clock and calling out thirty-second intervals. Students not watching the clock can record the beetles' position in relation to each other every thirty seconds, with two choices: (1) the two beetles are in the same grid square (at least half of each beetle's body is in the shared square), or (2) the beetles are not in the same grid square or if they are, less than half

of one beetle's body is not in the shared square. Fifteen minutes of observations is enough—you'll have thirty observations at that time.

If you choose to have groups of students do this experiments in sequence with the same two bessbugs, then each group can do maybe three minutes of observations, one every thirty seconds—six observations in all. With five groups, you'll wind up with thirty observations.

During the timed intervals, students who aren't watching the clock should continue to jot down anything they notice about the beetles' behavior.

Be sure not to leave the bessbugs in the box for long periods. I wouldn't keep them away from their terrarium and moist wood for more than an hour at a stretch. They are adapted to a humid space and will dehydrate and die in a dry box.

Result:

In stating the results, students report their observations and their counts. In my experience, bessbugs on a grid move around independently of one another. But this could vary with individuals, and could vary depending on how long they've been away from their colony.

Conclusion:

Students either accept or reject their own hypotheses, then write whatever conclusions they may draw from the counts and observations. They try to make sense of whatever they observe.

Although bessbugs are colonial and do behave cooperatively inside the tunnels of their habitat, I've found that they are not likely to pay attention to each other in the stressful and unnatural conditions in an empty box. They are more likely to give attention to getting out of the box. I've noticed that, inside a box, beetles that encounter one another often react by trying to crawl under each other, as though the other was just an object.

I've also noticed that bessbugs' affinity for one another may decline with length of time in captivity. When I first acquired the two beetles I have now and put a big piece of a rotted tree branch in their terrarium, they chewed a tunnel in the wood and stayed in the tunnel together. Since that time, I've taken them out of the terrarium multiple times, and some of the times I've had to tear away pieces of their terrarium wood to get them out. When the wood gets too fragmented, I add a new large piece from the ground in my backyard. As time has passed, I've noticed that they no longer occupy the same space in the terrarium. When I took them out to check something for writing this chapter, they were 8 inches (20.3 cm) apart in the terrarium. One was in a tunnel, the other was just between two pieces of rotting wood. My take on that is that their social behavior has degenerated as time away from the colony goes on—they've been away for several months now. They may be disoriented by being in abnormal conditions. Anyway, it makes me sad! I can't wait to go put them back under the log where I found them.

Additional Resources for Bessbugs

Life history and how to care for them:

"Bess beetles." FossWeb. Lawrence Hall of Science, The Regents of the University of California, 2012, lhsfoss.org/fossweb/teachers/materials/plantanimal/bessbeetle.html.

Video on bessbugs and how to set up a bessbug habitat:

"Bug Chicks, The." *The daily antenna: Bess beetles.* Vimeo, vimeo.com/41445867.

Woody, Tim, for CarolinaBiological. "Critters in the classroom: Bessbugs." YouTube, www.youtube.com/watch?v=CWEBkpVl1vs.

Experimental design for how to test pulling strength:

Massengale, C. "Bess beetle lab." *Biology Junction*, www.biologyjunction.com/10sBessBeetleLab.doc.

Chapter 6
Madagascar Hissing Roaches: Simply Stunning

Introduction

These roaches are cool, surprising, and will grab attention in your classroom. They're easily one of the most interesting insects I've ever had the pleasure to meet. Students who hold and interact with a hissing roach will never forget the experience.

Don't be put off by the word *roach*—these are wingless rain-forest roaches, and their appearance and behavior is very different from any roaches you may have encountered domestically. I can vouch for that personally. I live in an area of the southern United States that has a large and obnoxious roach species that gets in houses (the smokybrown cockroach). Roach *pests* are among the few insects that I really don't like—along with mosquitoes and fleas. So I am by no means a roach fan in general.

Fig. 6.2 Adult male hissing roach, showing spiky legs. His head is tucked under his thorax and not visible. The front part of the visible body is the thorax. The segmented rear part is the abdomen.

Fig. 6.3 Adult female hissing roach; head tucked under thorax and not visible. The front part of the visible body is the thorax. The segmented rear part is the abdomen. Note that the legs are less conspicuous on the female.

Fig. 6.1 Nia confidently displays the roach. These insects make a big impression!

That may be one reason I like the Madagascar hissing roaches so much. They're very different from their annoying relatives! An adult hissing roach is about 2.5 inches (6.3 cm) long and 1 inch (2.5 cm) wide. You probably wouldn't recognize one as a roach unless you looked at it from the side or front, which would allow you to see the roach-like head. But the head is kept tucked under, so from above you can only see the wingless thorax and abdomen. The Madagascar roach is actually an attractive insect; we shouldn't let domestic roaches give the family a bad name. The vast majority of roach species have no interest in human dwellings but instead mind their own business in nature.

Fig.6.4 This female hissing roach's head is visible from this angle, although the head is usually tucked under the thorax.

How to Get Madagascar Hissing Roaches

Hissing roaches (*Gromphadorhina portentosa*) are originally from Madagascar, an island off the southeast coast of Africa, so you won't be able to collect them in nature. Since Madagascar has a tropical climate, captive roaches need to be maintained at room temperature, 65 to 80 degrees F (18 to 26 degrees C).

The only way I know to get these roaches is to order them online. You can find online vendors that sell them cheaply as nymphs (immature). Carolina Biological Supply (CBS) and other vendors offer adults, which are more expensive (see Appendix). CBS and, I imagine, all other North American vendors raise them in captivity (which is very easy) so they're not removing the insects from Madagascar. Individuals can live two or three years in captivity—I've kept individuals even longer than that.

Materials You'll Need

Cages and food for captive hissing roaches are described in the following section. In the latter part of this chapter, each experiment will list any extra materials you may need for that particular experiment.

How to House and Feed Madagascar Hissing Roaches

I'll give you directions, but you can look online for other recommendations about how to house and feed Madagascar hissing roaches. You'll easily find such websites; these guys are popular pets.

Here's how I do it: I keep my three males isolated in three separate terraria, each about 12 inches (30.5 cm) long and 6 (15.2 cm) to 8 inches (20.3 cm) wide. My two females reside together in a fourth terrarium. In their native habitat in Madagascar, these roaches eat decaying organic matter from the forest floor. I've covered the floor of my terraria with soil from outdoors, so the soil is mixed with leaf fragments and other plant detritus. On top of the soil I've put a couple of pieces of a dried-out and decomposing tree branch from my backyard. I also have in each terrarium some lightweight pieces of curved bark (6 inches long or more), stacked and crossed to create spaces between them (a few large, thick leaves could work if you can't find bark).

The roaches desperately want cover during daylight hours, and they prefer a space with a low ceiling. A 1-inch (2.5 cm) ceiling of bark is better than a 3-inch (7.6 cm) ceiling. Some enthusiasts say the roaches like paper towel rolls for cover, but mine don't. They do enjoy an empty and closed egg carton, although they sometimes wedge themselves head-first into the holes underneath the carton and are very difficult to extract!

Hissing roaches are scavengers in nature, so they'll eat most fruits and vegetables. I feed mine small amounts of apple, banana, shredded carrot, squash, and celery. It doesn't have to be crisp and fresh, but don't give them moldy or rotten stuff. I give them dry dog food or cat food too. You might soften it with a little water at first until they get used to it. I check the food and water every day, and replace it if it starts looking rotten or moldy. I usually place the food in a jar lid or a medicine bottle lid to keep it from dispersing all over the terrarium and attracting fruit flies. Since the roaches are nocturnal, I rarely see them eating. So it's hard to know what they prefer—that's why I offer a variety. I also keep in each terrarium a very small dish of water, made by cutting the bottom half-inch off a single-serving yogurt cup, with a couple of cotton balls in it to keep the roach from drowning.

Hissing roaches do not ever need to have their cage sprayed with a water mist. They drink from the water dish. Because of this, I purposely avoid misting the cage because it makes the food mold faster, and there is no reason to. I suspect it could also make the cage stink because these are large insects and the leaf litter, of course, over time accumulates their dry fecal pellets. I never notice these, but they're there nonetheless. I clean out each terrarium, replacing all the materials, about every three weeks (I transfer the roach temporarily to a jar or other container first).

I keep the male roaches in separate quarters because males that are habituated to other males lose some of their aggressive behaviors. That would reduce your chances of seeing these behaviors, which I'll describe in the following sections.

Females may reproduce—be careful! I keep my two females separate from the males because I don't want them to mate and reproduce. If you have a female and she does reproduce, you'll have twenty to forty young to deal with, and you can't let them go unless you live in Madagascar. Introduced species can cause huge problems with local ecology, including displacing local species. Please don't release animals you've ordered—even creatures from other parts of the United States. Don't let students talk you into releasing them, as you'd be setting a terrible example for them. The best option for disposing of nonnative insects you don't want is to pass them on to another teacher, or you can freeze them—for insects, freezing is humane and painless.

If your roaches reproduce, you'll also have to be careful not to release the young into your building. I've read that this could be a problem, and who wants to find out? So a female must be kept in a terrarium with an absolutely secure lid, not a lid with slits in it.

It's possible that she could arrive in the mail already loaded with eggs. How would you know? You

wouldn't. Young roaches could easily escape through the slits in the common plastic terraria that have snap-on lids. I keep my females in a terrarium with a snap-on lid, but I keep a piece of cloth between the lid and the main part of the terrarium. The cloth droops down over the sides a bit.

If you want to avoid the possibility of offspring altogether, specify that you only want males when you order your roaches—although females tend to be calmer and easier to handle… so that can be tempting!

Background Information on Hissing Roach Behavior

Hissing Roaches Are Live-Bearers

Even though I won't let my roaches reproduce again, it has happened to me before, and it was plenty interesting when it did. Hissing roaches are live-bearers, meaning that they give birth to live young. This is unusual in the insect world. Most insects and most animals in general (except mammals) lay eggs rather than giving birth. The hissing roaches do produce eggs, but the eggs hatch internally and the mother gives birth to a gaggle of little ones, called nymphs. Hissing roach nymphs look something like roly-polies (pill bugs) because of the conspicuous segmentation on the back of their abdomens and the similar size. As of this writing, there's a YouTube video of a hissing roach giving birth, called simply "Hissing Cockroach Birth." I'm not sure whether to recommend it! I will say, however, that it's excellent videography, and it is memorable. As you might expect, there are many more hissing roach videos out there—some better than others.

Unusual Maternal Care

In addition to being live-bearers, female hissing roaches are unusual mothers in another way: they provide maternal care for their young. The vast majority of insect species abandon their eggs and the subsequent hatchlings to fend for themselves. But a mother hissing roach will stand guard over her young to protect them from a threat—such as your hand in the terrarium. She extends her legs to raise her body, and the young gather underneath her, at least the ones that can fit. Those left out seek hiding places elsewhere. It's an interesting sight!

As the nymphs grow, they'll shed their outer skin or exoskeleton about six times over a period of six months or so, then they'll be sexually mature adults. I kept my young roaches just long enough to show the students and to observe their mother's unusual maternal care. Then I told the students I was taking them home… which I did. To the freezer! It didn't bother me to freeze them, when I thought about the havoc they could wreak as an introduced species outside—and when I thought about all the domestic roaches I have smacked with a rolled-up newspaper.

What Does the Hissing Mean?

Yes, the hissing is communication. And this species is the only roach that hisses; that makes it special! The sound is created by expelling air from two specialized spiracles on the sides of the abdomen. Males do most of the hissing as a warning sound to perceived threats (like you and your students) or to rival males. If you pick one up, he may hiss—if you're lucky. The hissing is a treat! How many insects do you know of that hiss? If another male roach comes too close, he may give a territorial hiss to warn the other guy away. Males also make a courtship hiss to woo a prospective mate. The courtship hiss is a bit softer and shorter than an aggressive hiss. Females may hiss when disturbed, but they don't have a territorial or courtship hiss.

If the roaches are housed together so that male-male encounters are frequent, or if they are handled often, they hiss less. So I try to minimize handling of them, and house them separately. Students are vastly entertained by the hiss, though. Luckily, a roach that is rarely

Fig. 6.5 *An adult male has two prominent knobs on the thorax, the front part of the body in this photo (head is not visible). The knobs are used for butting in aggressive conflicts. The right knob is clearly visible in this photo.*

handled or disturbed may get so sensitive to disturbance that he will hiss if you even put your hand in the terrarium. And I don't believe that a hand in the terrarium will dampen hissing and other aggressive behaviors the way handling or exposure to another male does. You can be sure that if you remove a rarely handled and isolated male from his terrarium to engage in one of the following experiments, he will hiss. In my experience, for a short while on that day, he will hiss every time you touch him or try to lift him.

Although I've often heard aggressive hisses, I've never heard the courtship hiss. These roaches are nocturnal and do their courtship at night, unobserved. That's too bad!

Territorial Behaviors and Aggressive Anatomy of Males

Male hissing roaches have a fascinating array of aggressive behaviors, in the same way that male crickets do (see Chapter 9). If an unfamiliar male intrudes into another's "territory," the two will first touch antennae. Then, to defend his territory, the resident roach may engage in "posturing," where he elevates his rear by straightening his back legs, curves up the tip of his wingless abdomen, and gives an aggressive hiss. He may also wag the end of his elevated abdomen back and forth, a behavior called wagging or abdomen thrashing. If he gets really ticked, he may butt the intruder repeatedly. For effective butting, a male has two large knobs on the front of his thorax (*visible on Figure 6.5*).

The charging makes me think of a male horned mammal (such as a mountain goat). The butted roach may flatten itself to the ground in submission, or he may be knocked off his perch—it's quite a spectacle!

A female lacks the knobs altogether; her black thorax is smooth. Looking for the conspicuous knobs is the easiest way to differentiate the sexes (only adult males have knobs). The males' antennae are a little thicker than the females', but that's not as easy to distinguish.

Sexes Behave Differently— So Do Individuals

Males tend to be more active than females, in my experience. Females may be calmer and more cooperative in some of the experiments (such as the racetrack and the mazes). Also, individuals can have different personalities. I have no idea why, unless it's related to age or having come from housing that varied in the amount of crowding (which can be stressful to any territorial animal). Also, handling the roaches a lot during one lesson or experiment (even moving them by slipping a piece of stiff paper underneath them) can make the males gradually more agitated and less cooperative in experiments such as the racetrack and the mazes. That's because the handling is distressing to them. Being out in the open and exposed to light may also be distressing to them, since their natural tendency is to hide in the dark. Handling has less effect on females, or perhaps no effect, depending on the individual. For those reasons, using a female can be easier in racetrack and maze experiments—although perhaps less entertaining!

Thigmotaxis and Phototaxis

If you've looked at Chapter 7 on roly-polies/pill bugs, you've read that positive thigmotaxis is an organism's urge to have contact with solid surfaces around its body. This includes not only the underside of the body resting against the ground but also the animal's back or the sides of its body. Many ground-dwelling insects have the urge to seek out tight spaces so that they are closely sheltered. This urge for contact has survival value—it keeps them less exposed and less vulnerable to predators.

As you know, insects (such as moths) may also be attracted to light, which is called positive phototaxis. Others are repelled by light, such as roly-polies. Lots of insects are negatively phototaxic or photophobic. Light can mean exposure to predators!

It's difficult to tell whether hissing roaches are negatively phototaxic or positively thigmotaxic or both. During the day, they definitely seek out close confinement in dark spaces. But which of those factors is attracting them: the closeness or the darkness? I know that many hissing-roach researchers construct towers of closely spaced horizontal boards, and the roaches seek out the spaces between the boards. This suggests positive thigmotaxis. Furthermore, if I have an egg carton leaning against the inner wall of a terrarium, a male will often position himself between the egg carton and the glass so that both are actually pressing on him. That certainly suggests positive thigmotaxis. Yet, at night, my solitary male roaches sometimes sit on top of the logs and bark in their terraria. This behavior seems to indicate that their hiding during the day is only an avoidance of light. So I really can't say, from my own personal experience, whether hissing roaches are positively thigmotaxic. This is a good question to pose to students. Can they think of a way to separate those two

influences definitively? I've made some suggestions in Experiment 3 that follows.

Self-Defense

Hissing roaches don't bite in self-defense, can't sting, can't fly or hop—so how can they protect themselves from predators in nature? Or from a human hand trying to pick them up? I notice just three strategies. One is hiding: staying in dark places, being nocturnal. Another is running, and they can be fast when distressed, although not nearly as fast as domestic roaches and never so fast that they get away. The third strategy is to use their spiny legs. When I pick up a male roach with a finger on either side of its body, it pushes my fingers with the sides of its spiky legs. (See legs on Figure 6.2.) Although not painful, the spikes on the legs are startling and a little uncomfortable. It's enough to make me not want to pick up the roach unless I need to, and then I just do it anyway—it's not a problem. But I often find myself scooping up the roach with a piece of paper thick enough to support its weight so that I can avoid the legs. I can imagine if an animal tried to eat the roach, the legs would be unpleasant inside a mouth. Still, I'm sure these roaches are tasty morsels to hundreds of Madagascar species. Predators probably pull the legs off before eating. A calm roach, an elderly roach, or a female roach is less likely to use the spiky-leg strategy. In fact, my own female roaches never do. An agitated young male is likely to. And male roaches often do get agitated when called upon to participate in experiments!

Because of the spiky legs, students may be reluctant to lift a male adult roach by gripping it with two fingers, although some will embrace the challenge. When I forewarned students about the male's leg behavior, Lucia immediately declared, "I'm totally lifting that roach!"

And she did! It's easier, though, to lift and move adult males by sliding a thick piece of paper under them. If students want to hold a roach, they can hold it in an open palm, or on a forearm or shirtsleeve. A roach that isn't agitated from too much activity will sit calmly on an arm. Alternatively, a female roach is easy to lift and doesn't object other than maybe arching her back, which is amusing in its uselessness. An arched back could be a signal to another roach, but it has no effect on a predator or a person.

What Do Those Antennae Do?

The roaches have very long antennae (see Figure 6.2). They can't see well because their heads are tucked under the thorax, so they use the antennae to detect changes in the environment. If I take the lid off of a roach's terrarium to change the food, the roach stays hidden under the bark but will often stick one antenna out, like a little periscope, to detect by scent what's going on. Antennae sometimes get broken off, so you may receive a roach that has stubby antennae.

Madagascar Hissing Roaches at School

Observations and Activities

Unless you have a bunch of nymphs (immature roaches), you probably won't have enough roaches for every student to have one. Or you may acquire several adults, enough for groups of students to take possession of an individual. But even one hissing roach per classroom is enough for fun and class experiments.

In anticipation of your roach's arrival, students can help you set up a terrarium. Together outdoors, you can collect soil, pieces of bark, and small chunks of decaying logs or stumps. Or assign a few students to bring in this material. Students can bring in small bits of the foods hissing roaches will eat (see the "How to House and Feed…" section).

Students can be startled by the size of the roach when it arrives in the mail, and if not prepared, they may recoil from it. If a mass aversion to the roach sets in, or if you show any aversion, that attitude can be contagious. For that reason, it's best to psych them up with positive anticipation before it arrives. You can talk about how impressed their friends and family will be to see them hold such a large insect, and propose taking a photo of each student with the roach. Our local nature museum hosts birthday parties, and each attendee gets a printed photo of himself or herself holding a snake from the museum. Kids love that. A roach photo is in the same spirit of showing off their fearlessness. It will most likely be the largest insect any of them have ever seen (excluding insects with very large wings such as butterflies or dragonflies).

Be sure to tell the students before the roach arrives that it cannot bite, sting, fly, or run fast enough to escape. Tell them that the roach is completely vulnerable, much smaller than they are, and will be afraid of the students (it will be). I often point out to students that a hissing roach, or any small creature, is "afraid" that the students are predators who will eat it. This is true, although it's all instinctive, not involving any imagination on the roach's part. This changes the students' perspective, and usually engenders a certain empathy and care-taking attitude toward the animal.

After the roach arrives, an observation session will pique their curiosity. Have the students sit in a circle, then place the roach on the floor in the center of the circle. Ask students to describe its colors and physical appearance (large, wingless, segmented abdomen, six spiky legs, long antennae, head tucked under, knobs on a male's thorax for butting, etc.). All insects have segmented abdomens, but the segmentation isn't visible in insects with wings. Ask them to speculate on the functions of various parts of the roach's anatomy. What are the long antennae for? Why do they have spiky legs? Why do the males have two knobs like shoulder pads on the thorax, but females do not?

Ask the students to observe its antennae while the roach rests in the middle of your circle. Then (or sometime later) scoop up the roach with a piece of stiff paper and carry it temporarily to an open window. In my experience, a roach next to an open window or door will wave its antennae around, apparently detecting new scents on the air.

After the students describe the roach's appearance and consider functions of its body parts, watch what the roach does. (Probably nothing, at least for a while.) What does the roach do if you or a volunteer strokes its back with a finger? It may hiss, arch its back, try to run, or do nothing. Then you might place an egg carton or other small potential refuge near the roach and have someone stroke it again, or poke it from the side with a finger, gently. As their eyesight is poor, it may not see the egg carton, in which case it may run until if comes to something to hide under (perhaps a student's knee or foot).

Discuss any behaviors you might have observed so far: hissing, running, moving antennae, arching the back, moving legs to dislodge gripping fingers.

Ask who wants to be first to have the roach placed on their knee or arm. They may want to have you demonstrate first. Scoop up the roach with a piece of stiff paper to deliver it to a volunteer. Most likely, the roach will rest calmly on an arm. Students may prefer a sleeve over their arm, as the legs can feel scratchy.

After your observation session, you may want to tell students about some of the behaviors they are unlikely to observe on their own, such as maternal care or bearing live young (see the topics in the section Background Information on Hissing Roach Behavior). Ask them to speculate about some of the behaviors that will be addressed in the experiments that follow. How fast do they think the roach can run? Faster than any other insect? Do they prefer to hide or be in the open? If the students have had a roach on their arm, they've probably noticed the clinging ability. Just how well can they cling? Well enough to climb? Climb what? How could that be tested? What good are those knobs that look like shoulder pads? Can a male use them to push? How could that be tested? And so on. Encourage the students to come up with their own questions, and consider which ones may be testable with an experiment. It's a great experience for students to make up and test their own hypotheses when possible.

And now for experiments I have done with students.

Experiments for Madagascar Hissing Roaches

Some of these experiments are designed to explore the hissing roach's niche, or its relationship with its native environment. These experiments highlight the characteristics of the hissing roach that help it survive as a wild resident of the leaf litter on the forest floor in its native Madagascar. And some of the experiments are designed to provide a fun way to test hypotheses.

Remember that the hypotheses I give are just examples. Your hypotheses will be the predictions made by the class or individual students. Your result for each experiment will be a statement of how your animals reacted to your experimental setup. Your conclusion is a statement of whether your prediction was confirmed or not. For each experiment, adding replicates increases your confidence in the validity of your conclusion, but they may be omitted if tedious for young children.

Experiment 1

When I do workshops for science teachers or elementary teachers, I always include racetracks, because they're a student favorite, and they easily generate numbers to compare for testing hypotheses. Here, you'll find instructions for making a racetrack, and for timing a hissing roach on the racetrack. In Chapter 5 on bessbugs, you'll find instructions for racing two species at once on the racetrack, as a competition.

Question:

How fast can hissing roaches run on a racetrack?

Hypothesis:

Each student can predict how much time the hissing roach will take to move from start to finish on the racetrack described in the following section. For example, one student's hypothesis could be stated like this: "If we let our roach run the race-track ten times, I predict it will finish the race in an average of five seconds."

Materials:

1. One or more racetracks (see Methods, below, for instructions on making a simple paper racetrack).

2. One stopwatch or watch per racetrack (or students can use a classroom clock).

3. One hissing roach.

4. A clear cup or container of some kind to put upside down over the roach in the center circle to keep it from leaving the circle while the students prepare.

5. A weight to hold the cup down.

Methods:

The racetrack is just paper. Making it is easy, and once you have it, you can use it indefinitely, for any kind of small creature. (The racetrack is also used in Experiments 1 and 2 of Chapter 5 on bessbugs.)

The concept:

The idea is to make a circular racetrack, where the insect begins in the center and runs to the outer

Fig.6.6 A male roach streaks from the center circle to the finish line on this laminated racetrack.

circle (the finish line). The advantage of having this circular configuration is that the creature can head in any direction from the center; all directions eventually lead to the outside of the circle. Most creatures on the racetrack will head pretty quickly to the outer circle, in an effort to seek shelter or just from a natural inclination to bolt.

You can make this simple racetrack by drawing three concentric circles on a piece of paper. I use poster paper, 22 inches (55.8 cm) by 28 inches (71.1 cm). A sheet of art paper or newsprint about that size could work too, especially if laminated. The smallest inner circle should be big enough that the roach to be raced can fit inside of it—that will be the "starting gate." If you plan to use it later for races between two species (as in Experiments 1 and 2 of the bessbug chapter), then the smallest inner circle needs to be big enough for two insects. On my racetracks, the inner circle is 4 inches (10 cm) in diameter. The middle circle really serves no function other than for entertainment value, as in, "My bug crossed the middle circle first!" On my tracks, the middle circle is 10.5 inches (26.7 cm) in diameter, and my outer circle is 18 inches (45.7 cm) in diameter. The distance from the perimeter of the inner circle to the outer circle is 7 inches (17.8 cm). So in the race, I start the clock when the roach leaves the "starting gate" by crossing the perimeter of the inner circle, and I stop the clock when it crosses the perimeter of the outer circle—a distance of 7 inches. (In Experiments 1 and 2 of the bessbug chapter, I give approximate time to completion for various species that I've raced.)

How to draw those circles and make the racetrack:
   To draw circles that are actually circular, it helps to find circular objects that you can trace. I use pencil first, then when all three circles are good, I go over the pencil with a non-smearing, non-water-soluble felt-tip pen, such as a Sharpie. In making my first racetrack, I drew the biggest circle first (on poster paper). For tracing, I found a plastic trash bin at school that was almost 18 inches in diameter and centered it on the paper, just by eyeing it. Other possibilities for tracing the outer circle include trash-can lids, large pizza pans, big lampshades, or hatboxes. To trace the middle circle, I found a cooking-pot lid that was 10.5 inches (26.7 cm) in diameter. A plate might work. I placed the pot lid

within the big circle and measured the distance between its edges and the big circle to try to get it centered. After I traced around the pot lid, I found a coffee mug that was almost 4 inches (10 cm) in diameter to use for the inner circle. I measured from at least three points along its edges to the middle circle, so that the small circle would be centered within the middle circle. A jar lid, coaster, plastic cup, or a student's compass might work too for the inner circle.

I'm probably more meticulous than necessary in making sure all three circles are centered in relation to each other; it just depends on how much time you have. Doing it freehand is okay if you need to make it fast!

Fig. 6.7 Sophie watches the roach run the racetrack.

To avoid having to find that trash bin over and over for the big circle on additional racetracks, I traced the trash bin again on second piece of poster paper and cut out that circle to use as a template for making more racetracks. I have easy access to the cooking-pot lid and the cup, so I didn't need templates for drawing more middle and inner circles—but you could make templates for those too.

Of course, you can make your own circles any size you like. Use what you can find for tracing. I just used what was handy.

I've used the racetrack with a lot of different animals—vertebrates and invertebrates. Your racetracks will definitely last longer if you laminate them. My laminated tracks have lasted for years, and my husband uses them too for his classes. But you may want to leave a couple of racetracks unlaminated. If you plan to race other species (as described in Experiments 1 and 2 of the bessbug chapter), you'll find that some insects have trouble getting traction on the slippery lamination. Hissing roaches, however, can grip anything. Lamination makes no difference.

To race the roach:

In classrooms, I've had students work in groups of four students, with one racetrack per group. One student in the group can operate the stopwatch or watch the clock, one can record results, one can release the roach when everyone's ready, and one can retrieve the roach at the end.

If you have only one roach, this can be done as a class activity, or student groups can do the racetrack experiment in sequence.

When it's time to race, trap the roach under a clear plastic cup placed inside the inner circle of the racetrack, not overlapping the line. (See Figure 6.8.) If you leave a male roach under the cup for a few minutes, you'll need to either put a weight on top of the cup or have a student hold it down. Otherwise, he will easily push the cup around from inside. Females, in my experience, do not push the cup.

If you leave the roach under the cup for ten minutes or so, he may relax to the point that he doesn't move when you lift the cup. In this case, you can start the timer then anyway, or you can prod him or stroke his back to get him rolling. If you prod him, start the timer when he crosses the perimeter of the inner circle.

The results are more meaningful if you can do a number of replicates or trials (have the hissing roach run the race several times). Ten repetitions is optimal, but even three is better than one. As part of scientific methodology, decide before you start how many trials you're going to do.

Result:

Students' results are a statement of their own observations. If they record the time to completion for ten solitary runs from start to finish, then they should average the ten time measurements and compare this average time to their predicted time to completion.

In my experience, an adult male hissing roach can consistently cover a distance of 7 inches (17.8 cm) from the inner circle to the outer circle in one second!

A female hissing roach may be much slower than males, although I don't know why. I'm guessing it's because female hissing roaches tend to be much calmer in general.

Conclusion:

Students who made a correct prediction can "accept" the hypothesis in the conclusion portion of the data sheet. If not, then they reject the hypothesis.

If two bugs competed, then students count the number of times each bug won to determine whether their prediction of the "winner" was correct, and whether to accept or reject the hypothesis.

Suggested variations:

As an alternative, compare the times of two roaches. You're more likely to find a difference if you race male versus female. You can race them simultaneously—or if they distract one another or get too

Fig. 6.8 A plastic cup over the center circle can keep the roach in place while students get ready.

active to manage—then race them separately and compare times. To distinguish two roaches of the same sex, put a tiny dot of correction fluid on the back of one of them. If you do race two roaches, then each student predicts the winner as his or her hypothesis. Each student's "result" is a statement of the winner, and his or her "conclusion" includes any speculation about why the winner won.

Yet another option is to race the roach against other kinds of insects. (See Experiment 2 of Chapter 5 on bessbugs for instructions and information about potential opponents.)

Experiment 2

Question:
Do roaches seek out low-lying cover when given the option?

Hypothesis:
Each student's hypothesis is whatever he or she predicts. For example: "If we offer a hissing roach a low-lying shelf versus a space in open air, I predict that the roach will spend most of its time in the open air." Even if the class collects data jointly on a single roach, have each student make a prediction. Making predictions creates more interest in the outcome.

Materials:
1. A shoebox.

2. A sheet of cardboard at least as wide as the shoebox and at least half as long,

3. Something to trim the cardboard.

Methods:
To test this, you can easily create a crevice that the roach could crawl into by making a horizontal shelf just above the floor of a shoebox. (A square box will do; shape doesn't matter.) Cut a piece of cardboard the same dimensions as one end (one-half) of the floor of the shoebox. If the box is a typical shoebox shape, then your piece of cardboard will be roughly square in shape. (For example, if the floor of the shoebox is 7 inches 12 inches (17.8 cm x 30.5 cm), your piece of cardboard will be 7 inches by 6 inches (15.2 cm). Wedge the cardboard horizontally into one end of the shoebox so that it creates a shelf that is parallel

to the floor and uniformly about ¾ inch above the floor of the shoebox.

Next put a single roach into the box and cover the lid snugly with plastic wrap or a mesh you can see through, held in place by two tight rubber bands. The roach will have a choice of crawling under the shelf or staying out in the open. Give the roach at least a half hour to settle down from having been moved into the shoebox before you begin recording observations.

To collect data and test students' predictions, have individual students check and record the roach's position every five minutes or every ten minutes for one to three hours, depending on your schedule—the longer period if possible. Only one student is needed for any single observation. Settle on the timing before you begin, as part of the scientific method. This avoids the subjectivity of stopping when results are as you expected. If you need to take a break during testing, that doesn't matter. Just resume when you get back.

You need only one sheet to record data, placed beside the shoebox. Make the data sheet by drawing a line down the center of a piece of paper, creating two columns. At the top of one column, write: "Roach Is under the Shelf." At the top of the other column, write: "Roach Is Not under the Shelf." If the roach's head is under the shelf, but the rest of the body is not, I count that as being under the shelf because its sensory organs are under the shelf.

When it's time to begin, you can set a timer for ten minutes, and when it goes off, have one student check the roach's position and record the current time on the clock in the appropriate column. So if you do it every ten minutes for two hours, at the end you'll have twelve clock times written on the data sheet. For example, the "under the shelf" column might list four times the roach was observed as being under the shelf: 9:10, 9:20, 9:50, 10:10, in which case, students would have jotted down eight times the roach was observed as being not under the shelf.

If you have to interrupt and later resume the observations, don't leave the roach in the box for a prolonged period without a little bottle cap of water and some food.

Suggested variations:

1. Instead of having students write down the times, you can just have the students make a check in the appropriate column at the time they make their observation. It's nice to have them write down a time, though, because then you can see if the roach went back and forth a bunch or crawled under the shelf after a while and stayed there.

2. Of course, if you have more than one roach, the results will be more meaningful if you test them all. You can test them separately exactly as above, then average or total the number of times any of the roaches were observed under the shelf. Likewise, average or total the number of times any of the roaches were observed not under the shelf, and compare the two counts to see if they have a preference for staying under the shelf.

3. You could also put several roaches in the shoebox at one time, and then students would just tally the number of roaches out in the open at any given observation time. By subtraction, the observing student could also deduce and record the number of roaches under the shelf and record that. This option is fine, and easier. The only disadvantage is that you get no information about particular individuals changing position. But if they all stay under the shelf, that point is irrelevant. Anyway, frequency of changing position is not part of the hypothesis, so it is certainly not essential.

Result:

Your students' results will be a statement of their observations and the counts. In my experience, the roach will probably crawl under the cardboard shelf and stay there. It may spend a little time initially exploring the box for escape options but will eventually move under the shelf. Even if the roach's back (its dorsal side) is not in contact with the underside of the shelf, its antennae will sense the presence of the shelf just above it. However, roaches can have different personalities. One of my male roaches, Brody, often stands on top of the bark and wood in his terrarium during the day. I don't know why. Is it a coincidence that he's my latest acquisition, and thus has been more recently stressed by crowded conditions in the vendor's cages? The others stay hidden all day.

Conclusion:

The natural habitat or niche of hissing roaches is under leaf litter on the forest floor. They are nocturnal, so during the day they stay hidden under the leaf litter, or inside another dark place such as a knothole of a tree or a crevice in a stump. Since they are wingless and unable to bite or sting, they are extremely vulnerable to predators (and are sometimes raised as food for exotic pets in the United States). Staying hidden during the day and coming out only at night protects them somewhat from predation.

Experiment 3

Experiment 2 did not really determine why the roach tends to stay under the shelf (if it did). Is it avoiding light (negative phototaxis)? Is it seeking contact with shelter overhead (positive thigmotaxis)? It's one of the other; insects don't have reasoning ability to think through the advantages of staying hidden.

Here's a good way to answer this question of phototaxis versus thigmotaxis.

Question:

Does our roach spend more/less/equal time under the shelf in a completely dark room (as opposed to a well-lit room)?

Hypothesis:

We think our roach will spend more/less/equal time under the shelf.

Materials:

1. A shoebox.

2. A sheet of cardboard at least as wide as the shoebox and at least half as long.

3. Something to trim the cardboard.

4. A dark closet or a very thick dark cloth to drape over the shoebox, or some other light-tight covering. If you use a cloth, put the shoebox under a chair and drape the cloth over the chair. This allows students to lift the cloth to observe the roach without jarring or moving the box in any way. A dark closet may be easiest, if it is indeed almost pitch-black, because you can

open the door and quickly look at the roach's position before it has a chance to move.

Methods:

Use the same shoebox with a shelf that you created for Experiment 2. Again, put a roach in it.

Create a data sheet exactly as in Experiment 2, but write at the very top that the experiment was conducted in the dark. Before you start, determine the number of observations students will make and at what intervals. It's okay if you make fewer observations in the dark than you did in light, depending on how cumbersome it is to access the dark place. Try to make at least five observations at ten-minute intervals.

Result:

Your students will state their observations and the counts. I have not actually done this "shoebox with a shelf in the dark" experiment! Since hissing roaches are nocturnal, I'm guessing they will be more frequently observed away from the shelf when the box is in the dark. However, I wouldn't be surprised if there was no difference between the dark closet and the light room. Roach behavior is complicated. I have seen evidence of positive thigmotaxis (preferring contact with a surface overhead), which could affect their choice in light or dark. I have also seen roaches' resting spots influenced by the presence of other roaches.

I'm thinking too that after the roach has explored the box in the dark and found nothing of interest, and no escape route, it has little reason to stay out in the open. Whereas in nature, a roach has good reason to go exploring at night: looking for food and looking for mates. I want to do this experiment because I'm curious about the results!

Conclusion:

Your students will reject or accept their own hypotheses, based on the class's results.

Experiment 4

Question:

Can hissing roaches climb or cling to some surfaces better than others? (See Suggested Variation at the end of this experiment.)

Hypothesis:

Each student's hypothesis is a statement of his or her prediction about which surface hissing roaches can cling to best. A second hypothesis might be a statement of which other invertebrate might cling better or worse than a hissing roach.

Materials:

Try a variety of surfaces, ranging from slick to rough. These could include a ceramic plate, a piece of cardboard, a textbook, a plastic plate, a laminated sheet, the front of a binder, a file folder, a piece of unfinished wood, a piece of cloth taped to a flat surface like a book (not terry cloth—loops can catch and damage insect legs)—or any other flat movable thing the students can suggest.

Methods:

In my experience, this is most easily done as a group.

To create a data sheet, have each student write three words on a single line at the top of a sheet: "Surface," "Prediction," "Result." They will be generating a list under each of these words, so have them space the words accordingly. Under "Surface," have each student make a vertical list of all the surfaces to be tested. Under "Prediction," have each student write "yes" or "no" beside each surface to indicate his or her prediction of whether the roach will be able to cling to it or not when the surface is made vertical. Then have them select the one surface they think the roach will cling to best, and the one it will have the most trouble clinging to, and write "best" and "worst" beside those two, respectively.

Assemble on a table or on the floor all the surfaces to be tested. I find it easiest to collect the roach by picking up whatever object it's clinging to in its terrarium and nudging it from there into a plastic cup. Or you can slide a piece of stiff paper under it to transfer it to a cup. If it's a female, you can easily pick it up and put it in a cup. If it's a male and the spiky-leg jabbing bothers you, you can pick it up with a glove.

To test the roach, it's easiest to put it onto the surface while the surface is horizontal; you can just slide him onto the surface from the cup. Then hold the surface vertically over something soft to cushion the fall, in case the roach does fall off. Does it cling easily? Does it slide a little, struggling to get a grip? Or does it slide right off? Have students record their observations for each surface under the word "Result." They can write whatever you like. For me, these three choices cover all possibilities well enough.

Result:

In my experience, a hissing roach can cling to absolutely anything, including glass. I have never observed a hissing roach to lose traction, even for one second.

Conclusion:

Students are surprised that the roach can cling so well. Hissing roaches have tiny microscopic bundles of hairlike structures on the ends of their legs that somehow create suction. I don't see that such structures are needed for these denizens of the forest floor in Madagascar, but I think it's a characteristic of the roach family in general.

Roaches are famous for being able to swing around the edge of a table, from the top surface to the bottom surface, with only the rear two legs attached. This allows them to disappear almost instantly and certainly contributes to their success worldwide, because it makes them very difficult to pursue and catch. I've read that hissing roaches can cling to surfaces even better than geckos (lizards). I've been amused to see geckos running around upside down on smooth ceilings in Latin America. Especially in outdoor cafés at night, geckos cling to ceilings so that they can catch insects that come to the lights.

Suggested variations:

To heighten the surprise at the result of this experiment, it's fun to test one or two less clingy bugs first, using an identical data sheet, identical surfaces, and identical procedure as for the roaches. If you have mealworm beetles or bessbugs, those are ideal candidates because they can cling to almost nothing other than cloth or rough wood.

A fun wrap-up is testing a slug or snail. Because of their slime, they can cling to slick surfaces but don't cling well to cloth. Their mechanism of clinging (slime) is very different from the mechanism employed by hissing roaches (tufts of densely packed "hairs"). You might ask students how the two methods differ.

Experiment 5

Question:

How strong are hissing roaches?

Hypothesis:

Hissing roaches are strong enough to push 25 grams (or whatever each student thinks).

Materials:

1. A transparent plastic cup, with a rim diameter bigger than the hissing roach's total length.

2. A series of small weights, such as those in a student metric-weight set. The weights should range from 5 grams to 100 grams, in increments of 10 grams or so.

Method:

On a table where everyone can see, trap an adult male hissing roach under the inverted transparent cup, so that the rim of the cup is against the table and the roach is comfortably restricted within that space. Put a 5-gram weight on top of the cup and wait for the roach to push the side of the cup. If he doesn't begin pushing the cup within minutes, slide the cup gently onto a laminated poster; the cup slides more easily on the slick plastic. After the roach begins pushing, substitute a heavier weight and wait for him to push again. Continue increasing the weight on the cup in small increments until the roach is no longer able to push the cup.

Fig. 6.9 An adult male roach can easily push a cup with weights on it, using the knobs on his thorax.

Result:

The students record whatever results they observe. In my experience, a male roach can push a cup with 80 grams on top of it, on a laminated surface—but no more than that. On a paper surface, pushing is more difficult due to greater friction. The last time I did this, the roach began to struggle on a paper surface with 30 to 40 grams, and stalled out at 40 grams, unable to move it. But he continued to try over the next few minutes. At some point the pushing and the difficulty triggered an aggressive instinct in him, which was fascinating! He began hissing and wagging the end of his abdomen, territorial and aggressive behaviors that are typically directed at intruding rival males. He then began butting the side of the cup (hissing roaches do butt rival males) and was able to push it with 80 grams on it! He couldn't go beyond that weight, in spite of our urging him on.

Conclusion:

The conclusion is that male roaches can indeed push, as the knobs on a male's thorax are structured to do. Eighty grams is almost 2.8 ounces, pretty good for an insect. I can't think of any other insect that might achieve that. The accomplishment is a reflection of at least three factors: (1) the "shoulder pads" on the thorax, (2) the extremely good grip these roaches have for pushing against the floor, and (3) the butting instinct of male hissing roaches. Apparently, their territorial aggression can be triggered or enhanced by anything that resists being butted out of the way, including a plastic cup.

Students can feel the push:

Students can feel the pushing strength of a male roach with this simple procedure. Separate the bottom half of a cardboard egg carton from the top half and throw away the top. Locate the row of five "cones" along the midline of the carton. Make a fingertip-sized hole in the top of one of these cones, but leave the cardboard tip attached so that you can cover the hole as needed. With the hole covered, let a male roach crawl up into the underside of the cone that has a hole in the top. (He'll love it.) Once he's in place, uncover the hole. Now anyone can stick their finger into the hole and push gently against the front of the roach's body. You'll be pushing the front of his thorax with the "shoulder pads," because the head is tucked

Fig.6.10 A gentle touch might elicit a hiss.

under. The roach will push back, and his strength is surprising!

Experiment 6

Question:

Do the roaches hiss consistently, or does their motivation to hiss fluctuate?

Hypothesis:

The roach will hiss (not hiss) each time we remove him from his terrarium and touch him.

Method:

The purpose of this is to see how frequency of handling affects a male roach's tendency to hiss. On each predetermined test day, simply take him out of the terrarium and handle him in some way. An easy way to do that is to pick up the object he's resting on in his terrarium, and put it on a table where everyone can see. Then have a student stroke his back slowly but firmly. If that doesn't evoke a hiss, put a finger on either side of his thorax and pull gently as though trying to lift him. (This is an action that some predators would take in trying to pick him up to eat him). You don't need to actually lift him, just gently try. A suggested schedule for doing this is Day 1, Day 14, Day 15, Day 16, Day 30. On days when you're not doing this, leave him alone entirely, other than checking his food and water every day.

Result:

Students record their own observations. In my experience, a male roach left in solitude for two weeks, with no handling, is likely to hiss when stroked or lifted. A roach handled on consecutive days grows less likely to hiss. A roach handled every day is very unlikely to hiss.

Conclusion:

Roaches become habituated to handling and grow less likely to regard it as alarming.

Experiment 7

Question:

Do roaches have a preference in a T-maze?

Hypothesis:

We think roaches will turn left and right equally in multiple runs through a T-maze.

Materials:

1. A sheet of cardboard at least 10 inches by 13.5 inches (25.4 cm x 34.3 cm).

2. Pieces of cardboard (or thin foam core) of the following dimensions:

   • two strips that are 1¼ inches by 9½ inches (3.2 cm x 24.1 cm) (will refer to as "A strips").

   • two strips that are 1¼ inches by 3½ inches (8.9 cm) (will refer to as "B strips").

   • one strip that is 1¼ inches by 9 inches (22.9 cm) (will refer to as "C strip").

3. A hot-glue gun or two-sided tape.

Method:

On the sheet of cardboard, you'll be constructing a T-shaped corridor using the five strips of cardboard. You'll glue the strips on edge to create the corridors. Before you glue anything, draw the T-shaped corridor on the cardboard because it's hard or impossible to pull hot glue off of cardboard without ripping it. Alternatively, use two-sided tape to attach the strips to the cardboard; it can easily be moved.

To give you the overall picture before you start gluing:
You'll be starting the roach at the bottom of the T. The roach will walk up the long stem of the T and then will turn either right or left at the top, to exit through one side of the T. Students will predict and record which way it turns, over multiple trials.

On my best T-maze for hissing roaches, made of the materials above, the bottom of the T begins 1¾ inches (4.5 cm) from the lower end of the cardboard. The two longest strips ("A strips") are glued parallel to each other, creating a long corridor 1¾ inches wide. The long corridor is parallel to the long axis of the sheet of cardboard. At the top of the two long strips, I've glued the two shortest strips ("B strips"), intersecting the long strips at right angles and extending out sideways to make the lower part of the T's crossbar. The corridor at the top is also 1¾ inches wide, so the "C strip" is glued near the top of the cardboard sheet, parallel to the "B strips" and 1¾ inches above them. I write the letters L and R on the cardboard at the respective exits from the T-maze, just to be sure students don't get left and right mixed up.

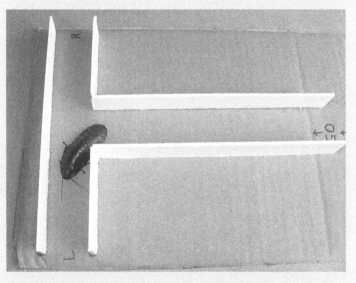

Fig.6.11 A female roach turns left in this trial on the T-maze. Note her use of her antennae in feeling for the sides.

I do this experiment in groups of four, so I don't have to make a T-maze for every individual. Alternatively, you can do it with one maze as a class experiment, or you can let student groups work in sequence, each group doing two to four trials. Each student should make his or her own prediction of the outcome. When students work in groups, I have each group do ten trials. On the data sheet I've created, there are two boxes for results, one to make tally marks for left turns, one for right-turn tallies. Underneath each box is a blank for their group total and another

Fig.6.12 Louie and Sophie watch as the hissing roach turns right in the T-maze.

blank for class totals. At the top of each data sheet, the question is already written ("Do hissing roaches have a preference for direction in a T-maze?"). Then there's a space for the student's hypothesis, followed by a space for students to write the procedure—something like: "We put our roach through the T-maze ten times and recorded its choice each time."

At the bottom of the sheet, students summarize results in writing, and then write a conclusion, accepting or rejecting their own hypothesis.

Depending on how gently they are handled, hissing roaches will in my experience eventually begin climbing over the walls of the maze. You can lay Plexiglas over the top, or tape plastic wrap over the top, but they can also walk upside down on the Plexiglas. This means your students may not be able to complete ten trials at one time. If that's the case, have them do fewer.

Result:

In my experience, roaches show no preference for left or right in a T-maze.

Conclusion:

Hissing roaches have no directional preference.

Experiment 8

Question:

Do hissing roaches have a directional preference in a T-maze if one of the exit corridors leads toward a light source?

Hypothesis:

We think our roach will not be influenced by having one exit corridor lead toward a light source.

Materials:

1. The T-maze you used for Experiment 7.
2. A light source. I used a door open to the outside on a sunny day.

Method:

Use the same procedure as in Experiment 7. The only difference will be that one of the short exit corridors of the T-maze (either left or right) will lead toward a source of light. An open door is ideal because then you there's no glare that could be objectionable. Run ten trials, or as many up to ten as your roach will tolerate.

Result:

Students state their observations and counts, recording the roach's choices on a data sheet as described for Experiment 7. In my experience, if one exit corridor (left or right) leads directly toward an open door, a hissing roach will consistently turn the other way, apparently turning away from the light. It's not guaranteed, of course, that yours will do this, but I expect it will.

Conclusion:

A T-maze can indicate that hissing roaches turn away from light. This makes sense, given that they are nocturnal and that most individuals hide under cover in darkness during daylight hours.

Experiment 9

Question:

Are male hissing roaches always aggressive to each other, or only in a terrarium where one male is living with a female?

Hypothesis:

Roaches are always aggressive toward each other.

Materials:

Two male and one female hissing roach.

Method:

Begin by isolating all three roaches in separate terraria for at least a week, then introduce one male (the "intruder") into the terrarium of the other male (the "resident"). Have students observe and record the reaction of both males over a period of at least five minutes, then return the intruding male to his own terrarium. Next, put the female into the terrarium of the resident male and leave her there for one or two weeks. At the end of that period, put the intruder male into the terrarium with the resident male and the female. Record and observe the behavior of both males. Be aware that the female might give birth after this. See pages 79–80 on how to keep nymphs from escaping.

Result:

The result is variable. Generally, males will not show aggressive behavior to each other if no female is present. Some researchers have found different reactions when a female is present and has been in the resident male's territory for some time. Some researchers have observed the resident male displaying a variety of aggressive behaviors toward the intruding male when defending a female. These behaviors can include hissing, posturing, tail wagging, and butting. (See page 80 for a description of these behaviors. You can see videos of the behaviors by searching YouTube for hissing roach aggression, hissing roach butting, etc.) I haven't been able to elicit these behaviors in my own roaches, although I have left my most active male alone with one and even two females for weeks at a time before introducing an intruder. I don't know why the resident has not reacted. It might be too much light, which motivates them to run for cover. My next step will be to try it in almost complete darkness, ideally using an infrared light. I know hissing roach behavior can and does vary among individuals. I hope you have success with this setup; I know others have.

Conclusion:

I will leave that up to you. My own conclusion is that the motivational state of any individual hissing roach is difficult to predict!

A different result:

See the paper "Territorial Behavior of Madagascar Hissing Cockroaches" by Justin Stine et al., at cockroachbehaviorproject.blogspot.com.

Additional Resources for Madagascar Hissing Roaches

General information about roach life cycle and habits, and suggestions for further reading:

Bullington, Stephen W. "Madagascar hissing cockroaches." Madagascar Hissing Cockroaches, users.usachoice.net/~swb/pet_arthropod/hiss.htm.

Clark, Debbie, and Donna Shanklin. "Madagascar hissing cockroaches." University of Kentucky College of Agriculture, www.ca.uky.edu/entomology/entfacts/ef014.asp.

Darmo, Lisa, PhD, and Fran Ludwig. "Madagascan giant hissing roaches." Carolina Biological Supply Company. Genetics/Living Zoology Department, Carolina Biological Supply Company and Lexington Public Schools, Lexington, Massachusetts, www.accessexcellence. org/RC/CT/roach.php.

Jessee, Ashley. "*Gromphadorhina portentosa*: Madagascan hissing cockroach." *Animal Diversity Web*. University of Michigan Museum of Zoology, animaldiversity.ummz.umich.edu/accounts/ Gromphadorhina_portentosa/.

McLeod, Lianne, DVM. "Madagascar hissing cockroach." Exotic Pets, exoticpets.about.com/cs/ insectsspiders/p/hissingroach.htm.

Interesting experiment on male hissing roach aggression, including step-by-step instructions:

Stine, Justine, and Danielle Hunsinger. "Territorial behavior of Madagascar hissing cockroaches." Retrieved April 28, 2011, cockroachbehaviorproject.blogspot.com/.

Video of adult male roach behavior:

Despins, Joseph. "Madagascar hissing cockroach male behavior." Retrieved January 19, 2009, www. youtube.com/watch?v=4aoF2HxJOKU.

Video of adult female roach behavior (an amazing video you'll want to share with squeamish friends!):

"Hissing cockroach birth—Pt. 1." Retrieved January 23, 2010, www.youtube.com/. watch?v=OhOGQINu0lk/.

Creatures Found In and Under Logs

Chapter 7
Roly-Poly or
Pill Bug Science

Introduction

Roly-polies' or pill bugs' claim to fame is the ability to roll into a tight little ball when disturbed. Both common names derive from this talent. The rolling up behavior alone is enough to charm most children.

Roly-polies are commonly assumed to be insects but are actually crustaceans. Because they have descended from aquatic animals, their bodies have not evolved to retain water very well. This chapter describes experiments to show how roly-polies reduce their water loss behaviorally. It also describes how to maintain a thriving population of roly-polies in a container indefinitely for exploration.

After you've done the experiments in this chapter and in Chapters 8, 9, and 10, refer to the table in Figure 1 in the Postscript, which provides a comparison of the results for the experiments in these chapters. Figure 2 in the Postscript is a blank table that can be filled out with the results of your experiments, for comparison.

Materials

1. Roly-polies or sow bugs. (Sow bugs are similar but don't roll up.) These can be found easily under stones or logs, or they can be ordered (see Appendix).

2. Petri dishes and lids. Three-and-one-half-inch (9 cm) diameter dishes work well. These can be bought at a science hobby shop or ordered by mail (see Appendix). One dish per child or per group is enough. Kitchen saucers with plastic-wrap lids will do as well.

3. Filter paper or paper towel circles. You need one sheet per dish, the same diameter as the dish.

4. One small apple or potato chunk—approximately $1/3$ inch (.8 cm) square—for each dish.

5. Enough sand to cover the bottom of one side of each dish for Experiment 1.

6. A sheet of dark paper (construction paper works well) for every two dishes for Experiment 2.

7. Two fruit fly vials (see Chapter 16) or Play-Doh or something else for making crevices or tunnels for Experiment 4.

8. Fabrics, paper, soil, etc., for offering different textures for Experiment 5. You'll need a container larger than a petri dish if you use vials to make tunnels.

9. Cups to hold roly-polies while experimental conditions in the dish are being altered.

Background Information

My young son loves to find roly-polies around the edges of the sidewalk and show them to anyone who'll look. He thinks there's something magical about them because

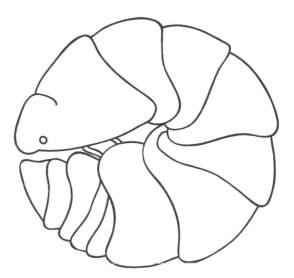

Fig. 7.1 Roly-poly rolled up.

of the way they roll into balls. Roly-polies are one of the easiest animals in this book to find, so most children will be familiar with them. On log-flipping expeditions, they are one of the first animals to be grabbed because they look so harmless.

Rolling into a ball is the pill bug's only way of protecting itself from predators (see Figure 7.1). I've watched praying mantises try to eat them and fail because their jaws apparently couldn't penetrate the rolled-up armor. *Armadillidium* is the genus name of the common roly-poly in my area, and that seems appropriate. The rolling up is interesting, but not nearly as interesting to me as the array of behaviors that roly-polies have evolved to cope with the problem of dehydration.

Dehydration

Life began in the ocean, and so the life processes that keep us all alive developed in a watery environment. All living cells are full of water that is essential to their functioning. Because air has a drying effect, keeping cells full of water is one of the major challenges of land animals. Insects were among the first invaders of land, so they have had time to become finely tuned for life on land. They've evolved a number of physiological adaptations for preserving their bodies' water content. And insects have consequently invaded every corner of the Earth, including the driest deserts. Efficiency of water conservation is not the only reason insects have been so successful, but it's one of them.

Roly-polies are not insects but rather crustaceans. Most crustaceans (crabs, lobsters, crayfish, etc.) are aquatic. The roly-polies' move onto land has been recent, in evolutionary time, so they have not had time to evolve the complex physiological adaptations to a dry environment that insects have—adaptations that involve the respiratory system, excretory system, and more. For example, insects breathe by means of a system of internal tubules that run throughout the body. The tubules can be closed to reduce water loss. In contrast, roly-polies have gill-like breathing structures that must stay moist and that lose a lot of water by evaporation. Insects also have a waxy covering that reduces water loss through the skin. Roly-polies lack this waxy cuticle and constantly lose water through the skin, especially when the humidity is low. In the animal kingdom, behavior is often the first line of adaptation to a new environment. Since roly-polies lack physiological adaptations for conserving water, they compensate with behavioral adaptations. That is the subject of most of the experiments I do with roly-polies,

which show how roly-polies cope with one of the major challenges of living on land.

How to Get and Keep Roly-Polies

Roly-polies are abundant in most areas of the United States. I find dozens when I look under the edges of the black plastic liner to my tiny backyard pond. I find them under bricks, under stones, under logs, under leaves, in the cracks between the cement porch and the brick wall of my house, or under a sheet of black plastic over my compost pile. I can collect fifty in a half hour without leaving my yard. I find sow bugs, which look similar but don't roll up, in the same places. They are also crustaceans, though, and their behavior is similar, so they'll do.

Keeping a thriving population in captivity is even easier than finding them. I keep them in a plastic rectangular food-storage container about 10 inches (25.4 cm) long, with a floor of 1 inch (2.5 cm) of damp sand. I also keep some in large margarine containers with damp soil. A damp piece of rotting log covers much of the surface of the sand. I throw in an apple core or slice of potato every week or so. They eat fruit as well as rotting plant material, fungi, and seedlings. I use a plastic lid or wax paper and rubber band lid, with a few holes, to keep the dampness in. The roly-polies reproduce like crazy in the containers. They stay in all the nooks and crannies of the piece of wood, and many burrow under the sand. They live up to four years—I have oldsters as well as tiny youngsters. The young look like miniature adults (there is no larval stage), except they're lighter in color. As they age they darken to a brownish gray. Rummaging around in a container like this will keep a child entertained for quite a while; it seems to be teeming with life. I keep a few other rotten-log creatures in there to make it more amusing: crickets, millipedes, or a few snails. I find them in the same places I find roly-polies.

Field Hunt

A class field hunt for roly-polies is fun and easy. Let each child take a milk carton or other container for captives. A group of five or so works well for me, if I'm flipping the logs and the students are looking for the insects. In addition to logs, flip over bricks, boards, rocks—or anything. Look in cracks and crevices around sidewalks. Make the point that we should always leave logs, bricks, and so forth, in their original position so that the damp places will stay damp. Otherwise, all the creatures that live under the log will die, their habitat destroyed. Bad for the wild things, and bad for the teacher and students. Because on the next field hunt, there'll be nothing to find.

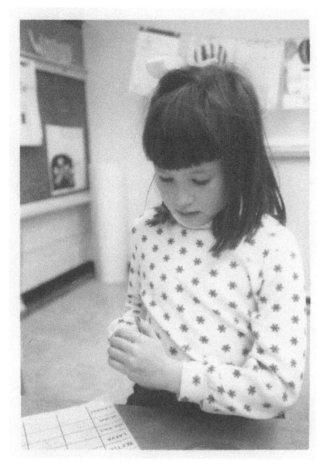

Fig. 7.2 Jennifer watches a roly-poly on her thumb.

Roly-Polies at School

Getting Ready

I use petri dishes for most experiments with roly-polies, although any small flat plate will do. Whatever you use must have enough of a rim to contain sand and to support a (transparent) lid. The dishes, the animals, and something to hold moisture are all you need to get started.

Observations and Activities

Before you do any experiments, you can get the children involved by letting them set up their own dishes. Give each student or group a dish and a piece of filter paper or paper towel cut to fit the bottom of the dish. Let them moisten the paper with a spray bottle or

by dripping water on it. Each dish will need a small piece of apple or potato.

Students enjoy choosing their own roly-polies from the big container, and most will enjoy letting the roly-polies crawl on their fingers (see Figure 7.2).

Occasionally a student will have an aversion to touching them or even looking at them. My experience is that if I respect the feelings of reluctant students and don't try to coax them into looking or touching, they often will come around on their own. It helps to let them see my and others' enthusiasm, and to give praise for any positive interest they may have. Tiffany, the little girl I described in the Introduction, was absolutely revolted by roly-polies at first. But after a morning of picking through rotten logs outside, she was the most ardent enthusiast of all. They say reformed smokers are the most zealous antismokers, and I guess the same logic applies here.

After each student has a roly-poly settled in a container, ask the students as a group to describe it to you. Then ask them to describe what the roly-poly is doing. Some will be running laps around the dish, some will be on the apple piece. If there are any blips along the edge of the filter paper, the roly-polies probably will crawl under them because they like to burrow.

While the students are watching is a good time to lead them to notice that roly-polies are not insects by asking them about the characteristics of an insect (six legs, three body parts) (see Figure 7.3). You can tell them then that rolies are crustaceans and ask them where other

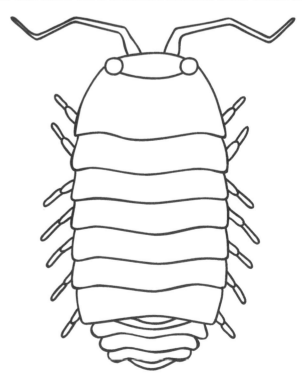

Fig. 7.3 Roly-poly at rest.

— 101 —

crustaceans live, like crayfish and lobsters, and crabs. (Most of them live in water.)

As noted previously, roly-polies are attracted to dampness, and moving about until they've found a damp spot is one of the behaviors that helps them survive in a terrestrial environment, in spite of inefficient physiological mechanisms for water retention. If you ask children to describe where they found the roly-polies, they'll say under bricks, in cracks, under a sheet of plastic, and so forth. Ask them what it's like under those places. They're unlikely to say "damp" at first because they haven't yet learned to think in terms of what features of the hiding places might be important to an animal. They're more likely to say yucky, hard, and dirty. Here's where you may need to lead them, by asking, "Was it damp or very dry?" Most will say damp. "Was it dark or light?" Dark. "Was it close and cramped, or roomy and airy?" Close and cramped. "So what kinds of places do you think roly-polies like?" Damp, dark, close and cramped (and maybe dirty, yucky, and hard too!).

At some point you may want to talk to the students about what features of any place are important to an animal. Most little animals are interested only in staying alive, and what they look for are features most likely to keep them alive. What do they need to stay alive? Food, water, and safety from predators. Aesthetics (yucky and dirty) are irrelevant to most animals. The risk of death is so great in small, vulnerable animals that avoiding death is really their only concern. A mouse or an insect that takes a stroll to find a cleaner and more attractive home is likely to get eaten or find itself with no home at all.

The first two experiments are designed to determine if the generalizations you and the students have made about what kinds of places roly-polies like are valid. Or is it something else that leads them to hang out under rocks? Maybe they're just hiding from predators or looking for supper.

Experiments

Remember that the hypotheses I give are just examples. Your hypotheses will be the predictions made by the class or a particular student. Your result for each experiment will be a statement of how your animals reacted to your experimental setup. Your conclusion is a statement of whether your prediction was confirmed or not. For each experiment, adding replicates increases your confidence in the validity of your conclusion, but they may be omitted if tedious for young children.

You've gotten the students so far to notice that all the places they found the roly-polies were damp. Now it's just a short jump to get them to propose the experiment itself. You could start by saying something like, "You children have said you think roly-polies may be attracted to damp places. What kind of choice could we offer the roly-polies to see if that's really true?" Someone may eventually suggest that you make one side of the dish damp and one side dry.

You can let each child who's interested do the experiment in his or her own dish, or do it as a class with one or several dishes.

Experiment 1

Question:

Are roly-polies attracted to dampness?

Hypothesis:

If we offer roly-polies a choice between damp and dry, they will choose damp.

Methods:

There are several ways to make a damp side and a dry side in each dish. You can use two semicircles of paper, one damp and one dry for each dish. Or you can use sand, dry on one side and damp on the other. Because very young children can set it up themselves, I use damp sand on one side and bare plastic on the other. Damp sand packed hard holds its shape very well. Each student can scoop the sand into his or her own dish and shape it into one half of the dish. The filter paper and apple or potato chunk that was in his or her dish can be set aside in a cup with the roly-poly while the sand is put in. You can save used plastic cups (e.g., yogurt cups) for this purpose.

One problem with using damp sand on one side and bare dish on the other is that you can't be sure whether the roly-polies are reacting to the dampness or to the texture of the sand. They do like to burrow. To control for this, you would ideally put dry sand on the other side. But then you have to deal with the problem of the dry sand wicking moisture from the damp sand. A strip of aluminum foil or plastic on edge between the two sides will help that. There's the same problem if you use wet and dry paper, but that can be solved with either a low barrier or a space to stop the wicking.

After the dish or dishes are set up, put a roly-poly in the center of each one. Put only one roly-poly in each dish because they influence one another's behavior when together. Make sure it can cross freely from one side to the other. Decide at the beginning how long you're going to wait to check the roly-polies' choice of sides. Give them several hours to settle down. The more roly-polies you test, the greater will be your confidence in your results.

Result:

The result is a description of the roly-poly's behavior during the experiment and its location (damp or dry) at the end of the experiment. You can use a histogram or bar graph to record your results (see Figure 7.4.) A blank graph is provided for your results (see Figure 7.5.)

Conclusion:

If offered a choice between damp sand and dry plastic, roly-polies will almost always choose the damp sand. The prediction is confirmed.

Expriment 2

Approach this one the same way you approached Experiment 1, by getting the students to state their observations about where you find roly-polies. If someone mentions dark places, then ask what kind of choice you might offer to see if they really do prefer dark places.

Each student who's interested can do the experiment in his or her own dish, or do it as a class.

Question:

Are roly-polies attracted to darkness?

Hypothesis:

If we offer roly-polies a choice between light and dark, they will choose dark.

Methods:

The bottom of the dish should be uniformly damp. Paper is probably preferable because the roly-poly may burrow in sand and thereby avoid the light. If you use dry paper you may get a quicker response to the light, but be aware that your roly-poly may dehydrate and die quickly without dampness in the dish.

I create darkness for the roly-poly by taping a piece of black construction paper over one half of the dish, enough to extend beyond the edge of the dish (see Figure 7.6). This shades the edge too, making the inside darker. Put the roly-poly in the middle of the dish, put the lid on, and leave it for a predetermined interval of several hours. If your roly-polies show no preference, try putting the dishes close to a lightbulb.

Result:

Your result is a statement of the behavior of the roly-polies in the dark/light dish. You can use a histogram or bar graph to record your results (see Figure 7.4). A blank graph is provided for your results (see Figure 7.5).

Conclusion:

If offered a choice between light and dark, most roly-polies choose dark. The prediction here is usually confirmed.

There are, however, several species of roly-polies and sow bugs, and some react to light differently from others.

It may not be obvious to a child that light, and the heat usually associated with it, can be drying. You may need to prompt this realization by asking what would happen to various items left out in the sun. Have they ever seen a dried-up earthworm on the sidewalk? How does a grape become a raisin?

You can do an experiment to determine whether roly-polies have a preference for warmth or coolness, unrelated to lighting. You can heat and cool opposite ends of a tray by following the procedure described in Chapter 8, Experiment 2, for earthworms. Roly-polies and most other crawlies prefer the cool end.

Have you ever huddled against another person when it's cold? Or crouched into a ball with your arms crossed tightly against your chest? We do this instinctively to reduce our surface area when it's cold. By making contact with other warm skin over at least part of our body's surface, we reduce the amount of heat that radiates from it.

Roly-polies use the same strategy to conserve moisture—huddling together works the same way to reduce water loss. By making contact with other

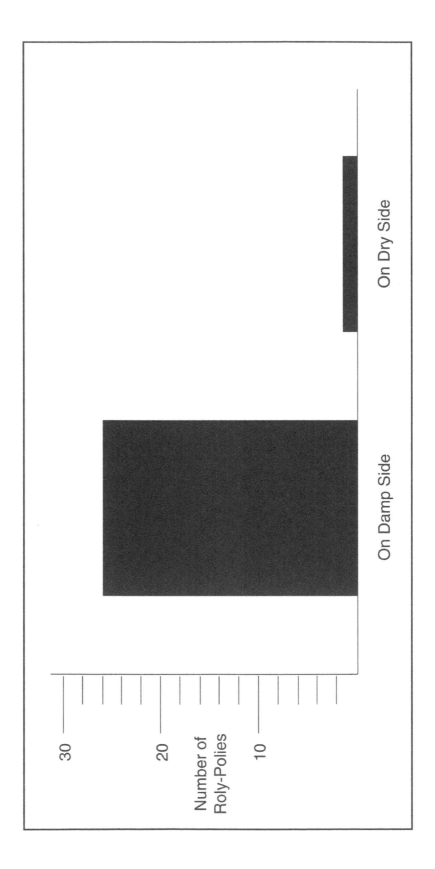

Fig. 7.4 Number of roly-polies choosing a damp substrate and a dry substrate.

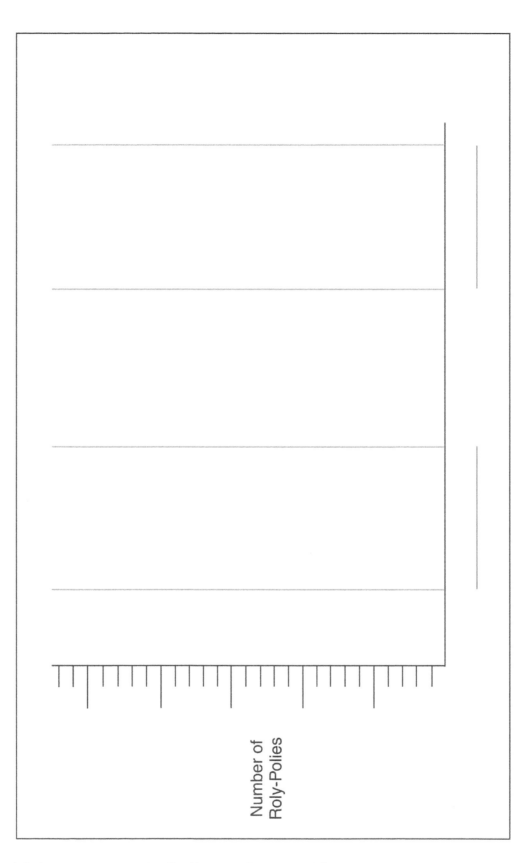

Number of Roly-Polies

Fig. 7.5 Blank graph for Experiments 1, 2 and 4. (See Figure 7.4 for an example.)
- *For Experiment 1, label the two histogram bars "On Damp Side" and "On Dry Side," as shown in the completed histogram (Figure 7.4). Number the vertical axis to match your sample size.*
- *For Experiment 2, label the two histogram bars "On Dark Side" and "On Light Side."*
- *For Experiment 4, label the two histogram bars "In Closed Space" and "In Open Space."*

Fig. 7.6 Claire peeking at her roly-poly on the dark side of the petri dish in Experiment 2.

roly-polies or really by making contact with any surface, roly-polies can reduce the amount of surface area exposed to the drying effects of air.

Experiment 3

Question:

Will roly-polies bunch to reduce water loss?

Hypothesis:

Roly-polies will bunch, or huddle together, when in a dry environment.

Methods:

You'll probably have to do this as a class, unless everyone has two dishes and several roly-polies. Set up one dish with a damp floor of sand or paper and another with a dry floor. Put several roly-polies (ten or so) into each dish. Wait for a predetermined interval (several hours) and then check their position. Are they in contact with each other (or very close) in either dish? If not, wait longer until the ones in the dry dish begin to dehydrate.

Result:

Your result is a statement of how each group of roly-polies behaved in response to their dish.

Conclusion:

Roly-polies in a dry dish will aggregate, or bunch, to reduce water loss, so the prediction is confirmed. Those in a damp dish may too, but not as much. Does it work? Does bunching reduce water loss? Children may find this unacceptably cruel, but one way to find out if bunching improves the survival of roly-polies is to compare the survival of several

that are allowed to bunch to the survival of one in isolation. The dishes should be equally dry. If bunching helps reduce water loss, then the several roly-polies together should survive longer than the one in isolation, which will probably die pretty quickly in a dry dish.

Experiment 4

Moths fly toward a light source. In the study of animal behavior, any innate tendency to move toward or away from a particular kind of stimulus has a name (as discussed in earlier chapters). For example, the tendency to move toward or away from light is called phototaxis. So moths are positively phototaxic. Movement away from light is negative phototaxis, or photophobia. Roly-polies, in addition to being positively hydrotaxic (attracted to moisture), are also positively thigmotaxic. A thigmotaxic animal will seek out a position where its body has contact with something besides air on all surfaces.

One of the first things a class of first graders noticed about their roly-polies in petri dishes, each with a damp piece of filter paper, was that many crawled under the filter paper if there was a wrinkle that made a small tunnel. Positive thigmotaxis could be responsible for that, although the space under the paper was probably more humid too. Thigmotaxis is responsible for bunching behavior. If no other roly-polies are available, how else might roly-polies satisfy their thigmotaxic urges?

Question:

If we offer a roly-poly a crevice or some other enclosed space, will the roly-poly seek out that spot? Is a roly-poly in a dry environment more likely to do so?

Hypothesis:

A roly-poly in a dry environment will seek out a crevice, while one in a damp environment may not.

Methods:

The students can volunteer their roly-polies for one of the two "treatments," damp and dry. In both types of dish, supply some sort of crevice the roly-polies can crawl into. You can make a trough of Play-Doh or clay or plastic, just wide enough for the roly-poly to get into, about 1 to 2 inches (2.5–5 cm) long. Leave it open at the top to be sure the attraction isn't darkness. Or, in a dish bigger than a petri dish, submerge a transparent fruit fly vial (see Chapter 16) horizontally in damp sand, so that the protruding side is only $\frac{1}{4}$ inch (.6 cm) off the sand. The inside of the vial should contain sand at the same level as outside the vial. This creates a long, low-roofed tunnel for the roly-poly, which is light inside, to rule out the effects of darkness. Set up another vial the same way in dry sand. Put one roly-poly in each dish and give them a predetermined amount of time to settle themselves (several hours).

To add replicates, repeat the procedure with more individual roly-polies, one at a time. Do any roly-polies go in the crevice or the tunnel? Are those in the dry dish more likely to do so than those in the damp dish? Do the roly-polies in the tunnel (fruit fly vial) stay where the roof slopes into the sand, where it is low enough to touch their backs?

Fig. 7.7 Craig ponders the meaning of his observations, trying to tie it all together.

Result:

Your result is a description of how the roly-polies responded to the setup in each dish. You can use a histogram or bar graph to record your results (see Figure 7.4). A blank graph is provided for your results (see Figure 7.5).

Conclusion:

If we offer roly-polies a crevice or a tunnel in a dry environment, they'll go in. They are thigmotaxic. If we offer roly-polies a crevice or tunnel in a damp environment, they may go in but not as often. They show less thigmotaxis in a damp environment.

Experiment 5

Question:

Do roly-polies prefer some textures over others? Do they prefer substrates that allow burrowing? Does humidity affect any preferences?

Hypothesis:

Roly-polies prefer substrates that are not perfectly flat and smooth but rather are bumpy and conform more to their bodies. They particularly like substrates that they can burrow into.

Methods:

Offer the roly-polies a variety of substrates in pairs: two different semicircles in the same petri dish. Offer damp choices and dry choices. You can try wax paper, unwaxed paper, wool, flannel, other fabrics, sand, soil, and so forth. The papers and fabrics may need to be taped or glued down; two-sided tape works well. Give the roly-polies time to settle down, and check their location. Can you make any generalizations?

Result:

Your result is a description and tally of your roly-polies' choices.

Conclusion:

Roly-polies tend to prefer irregular surfaces and avoid slick surfaces. They especially prefer substrates that allow burrowing. These preferences seem to be related to their thigmotaxis, and are less pronounced when humidity is high.

These experiments are all related. A table may help the students tie it all together. Figure 7.8 provides a sample table, with not necessarily "correct" answers—animals may vary. Figure 7.9 supplies a blank table on which to record your results. Your results may all show that rolies avoid the drying effect of air and heat any way that they can. Or you may have mixed results.

Experiments don't always turn out the way you think they will, which is in itself a valuable lesson about how progress is made in scientific research. You may have a species that responds differently from most of the others—for example, roly-polies

| Stimulus | Rolies' Choice | | Effect of Choice on Rolies | |
	Attraction	Avoidance	Drying	Keeping Damp
Dampness (Experiment 1)	✓			✓
Light (Experiment 2)		✓		✓
Heat (see Experiment 2, Chapter 4 for directions)		✓		✓
Other Rolies (Experiment 3)	✓			✓
Crevices (Experiment 4)	✓			✓
Rough Textures (Experiment 5)	✓			✓
Substances That Allow Burrowing (Experiment 5)	✓			✓

Fig. 7.8 A sample table summarizing the choices of the roly-polies.

Stimulus	Rolies' Choice		Effect of Choice on Rolies	
	Attraction	Avoidance	Drying	Keeping Damp
Dampness (Experiment 1)				
Light (Experiment 2)				
Heat (See Experiment 2, Chapter 4 for directions)				
Other Rolies (Experiment 3)				
Crevices (Experiment 4)				
Rough Textures (Experiment 5)				
Substances That Allow Burrowing (Experiment 5)				

Fig. 7.9 Table for recording results.

that don't respond negatively to light. The students can speculate about why you got the results you did (see Figure 7.7). They may conclude that we can't always explain our results, which is a fact of science.

It might be interesting to repeat one or all of these experiments with an insect that you know to be more tolerant of dry conditions—maybe a beetle or a small grasshopper. You may get some of the same results. My daughter compared the behavior of *Tenebrio* beetles (see Chapter 18) in dry dishes to that of roly-polies in dry dishes. To our surprise, she found that the beetles bunched too. She couldn't explain the results, but she won first prize for the second-grade science fair anyway just because she was the only one who did an experiment.

Chapter 8
Wiggly Earthworms

Introduction

Children love to hold slippery, slimy, wiggly worms, especially big nightcrawlers (see Figure 8.1). Both older and younger children enjoy predicting the worms' preferences in terms of living conditions. Do the worms like heat or cold, dampness or dryness, light or dark? Earthworms can be tied into a study of soil, our dependence on soil and plants, and the destruction of soil with human development. (Consider showing your students the DVD "Dirt! The Movie" for a very informative and entertaining story of the value and abuse of soil.) Earthworms have a tremendously beneficial effect on soil. Their tunnels aerate the soil and allow water to penetrate better. They mix the soil layers and increase the volume of soil. They improve soil texture and nutrient content. Several experiments explore these effects.

The maze (Experiment 9) is probably students' favorite exploration, other than just handling the worms. Can the worm find the right door?

After you have done the experiments in this chapter and Chapters 7, 8, and 9, refer to the table in Figure 1 in the Postscript, which provides a comparison of the results for the experiments in these chapters. Figure 2 in the Postscript is a blank table that can be filled out with the results of your experiments, for comparison.

Fig. 8.1 Felicia dangles a worm for inspection.

Materials

1. Earthworms can be bought at a fishing supply store, dug up at home, or ordered from a biological supply company. Other types of worms are available for order too, for the sake of comparison.

2. A tray and paper towels for Experiments 1 to 5.

3. A heating pad and bowl of ice for Experiment 2.

4. A lamp, dark paper, and a dish for Experiment 3.

5. A flashlight and red and blue (or other second color) cellophane for Experiment 4.

6. Dark loamy soil, like potting soil or good garden soil, for maintaining the worms.

7. Experiments 5 and 6 require sandy soil and clay soil or other soil types in addition to your loamy garden soil.

8. A worm-viewing cage for Experiment 6 (see Figure 8.2).

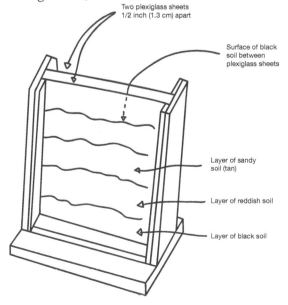

Fig. 8.2 Homemade transparent containers for seeing earthworms' tunnels.

9. A medium-sized jar and sand for Experiment 7 and several worm foods for Experiments 7 and 10.

10. Two potted plants for Experiment 8.

11. A Y-shaped tube for Experiment 9. You can order one from a biological supply company (see Appendix) or make one (see Figure 8.13).

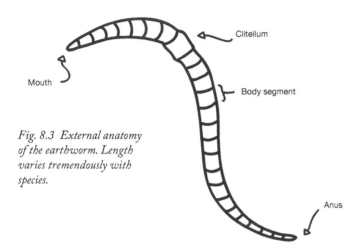

Fig. 8.3 External anatomy of the earthworm. Length varies tremendously with species.

Background Information

Earthworms in Relation to Other Worms

The word *worm* is used as sort of a catchall term for any long and thin and more or less legless creature that isn't a snake. Caterpillars are often called worms although they're really larval butterflies or moths. There are three large groups or phyla of real worms, which are only distantly related to each other. One group is the flatworms, which includes tapeworms. They're all essentially flat and are not all parasitic like the tapeworms. Land planarians are flatworms that I find under rocks in my yard occasionally. They're 4 to 6 inches (10.2–15.2 cm) long and have triangular heads. I can catch aquatic planarians, about $^3/_8$ inch (1 cm) long, by leaving a piece of raw meat in the few inches of water in a ditch behind my house. They look like a tiny black blot on the meat but will stretch out and swim around if you put them in a dish of pond water.

Another group or phylum is the roundworms. They're all round in cross section and smooth bodied. Some are parasitic; some are tiny and live in the soil (nematodes).

The third group, to which earthworms belong, is the segmented worms or annelids. They are round in cross section too, but their bodies are obviously ringed, or segmented. This distinguishes them from the roundworms. (You can order representatives of any of these three groups of worms, dead or alive.)

Lots of people react to worms with disgust, particularly the parasitic worms. But earthworms are completely harmless and clean, fascinating to children, and extremely helpful to humans through their effect on soil (see Earthworms and Soil section). I find that very few students, if any, react to worms with real distaste. Many are cautious and only want to look at first. They look to the teacher and others for cues as to how to react. With calm encouragement, but not pressure, most will want to hold the worms within five minutes.

I've used worms with younger students (kindergarten and first grade) as an introduction to a creature very different from ourselves and as an exploration of how its differences suit it. I like the students to see that even an animal as simple as a worm (no head, no legs, no obvious body parts of any kind) can have preferences and make decisions.

Earthworms and Soil

Earthworms can easily be tied into a study of the environment because of their benefits to soil. There are a number of ways to explore these effects through experiments. All earthworms create tunnels underground. Their tunnels help air and water penetrate the soil, which is helpful to plant roots. Their tunneling also mixes the different layers of soil, bringing minerals up from lower layers and taking the organic top layer down below.

Most important perhaps is that worms ingest dead leaves and other organic matter and convert it into nutrient-rich castings, which vastly improve both the texture and the nutrient content of the soil. Soil that is rich in castings holds together better and holds water better. The students can see the effects of the worms on the soil and how plants react to the improved soil.

I think it's good for children to see how humans can benefit from the normal activities of a creature as inconspicuous as the earthworm. This fosters an awareness of the interconnectedness of all creatures. Even our most advanced technology can't do for our soil and ultimately our vegetables what the little earthworm can do. By aerating the soil, mixing layers, and converting leaves into nutrient-rich castings, earthworms do more to improve the soil than all other soil animals put together, and soil is full of tiny organisms. We need worms!

The Body Parts of a Worm

The external anatomy of a worm is quite simple (see Figure 8.3). There is a "head" end, with a mouth at the tip. About one-third of the way from the head to the tail is a smooth band around the body, called the clitellum. I identify the head end by finding the clitellum, since it is closer to the head. The segments of the body are conspicuous. A segment is the space between two rings.

If you rub the worm gently between two fingers from the tail end toward the head, it will feel rough. You're feeling tiny hairlike projections called setae (pronounced see-tee) that help the worm hold on to the sides of the burrow.

Inside its body the worm has a long tubular digestive tract. Normally a worm likes leaves and similar organic matter to eat, but it can eat its way through soil that is too dense to push through. The leaves or soil are ground up in a muscular part of the digestive tube called the gizzard. The undigested part passes out the tail end as castings. The castings are technically feces but are not decayed or foul smelling. The castings are essentially just enriched soil.

A Worm's Unique Muscle System

The muscles of a worm are interesting for students who are old enough to understand that animals move by muscle action and that a muscle shortens when it contracts. If you handle a worm or watch it move across the tabletop, it's obvious that it can draw its body up to become short and thick. It does this by contracting the muscles that run the entire length of the body, under the skin. They're called longitudinal muscles.

But the worm can also extend its body to become long and thin. It does this by contracting muscles that encircle its body from head to tail. They are called circular muscles. The worm moves by contracting these two sets of muscles alternately, in different parts of its body. For example, it can extend the head end by contracting the circular muscles in the head end. (This increases the pressure in the fluid-filled body cavity, thus causing the squeezed segments to lengthen.) Then, while resting the head end on the ground, it draws the tail end up to meet the head end by contracting the longitudinal muscles in the tail end.

This squeeze-extend type of movement is very different from the movement of a snake, which lacks the sets of opposing muscles. Snakes move in several different ways, but generally by pushing against the ground with curves in their bodies. A snake's body does not change in either length or circumference as the worm's does. A movie of a snake or, better yet, a live snake would help students see the difference. Describing the locomotion of an earthworm in writing is a useful exercise in observation even for grown students. I find that very young children (five to six years old) have difficulty comprehending it.

An inchworm (really the caterpillar of a geometer moth), which has a very similar body shape, has a third type of movement. In some areas small green inchworms are very common and are found under oak trees. Look on outside walls or any structure off the ground under oak trees. Inchworms move by hitching up the rear to catch up with the front, but like snakes, they cannot change the length or circumference of their bodies.

Earthworm Gender—They're Hermaphrodites!

Each earthworm is both a male and a female; that is, each individual produces both egg and sperm. Being both sexes at once is called being a hermaphrodite and is not that uncommon in nature. Hermaphroditism can be an advantage. It's advantageous to earthworms because they're solitary and burrowing and don't meet other earthworms very often. If the sexes were separate, a female worm looking for a mate would have only a fifty-fifty chance that the next worm she encountered would be a male. Since they're hermaphrodites, earthworms have a 100 percent chance that the next mature and healthy worm encountered will be a suitable mate. When the worms mate, they align their bodies and exchange sperm simultaneously. Then each forms a gelatinous ring with its clitellum, the smooth band around the body near the head end. As it slips this ring off the end of its body, the eggs are deposited in it and the ring closes to become an egg sac.

How to Get and Keep Worms

In North Carolina I can dig up worms outside even in the middle of the winter, but in areas where the ground freezes they're accessible only part of the year. I have my best success in a garden or under leaf litter in the woods when the soil is damp, although roots get in the way in the woods. If you dig with a shovel you will chop many of your worms in half, fatally. A pitchfork causes fewer injuries. Or you can dig up a big hunk of soil with the shovel and break it apart with your fingers.

If you live in an area where fishing is a possibility, someone will probably have worms for sale. In the phone book, I can find worm suppliers listed in the Yellow Pages under "fish bait" and "fisherman supplies"; you can also search online. The local tackle shop sells three kinds of worms: redworms, Georgia wigglers, and night crawlers. These are different species of earthworms, varying in size from 2 to about 7 inches (5–17.8 cm) for a fully extended night crawler. Children find the night crawlers most interesting, although any will do. Ask the supplier which type lives longest in captivity and for tips

Fig. 8.4 Craig finds Jennifer's worm more interesting than his own. Note the long-term worm storage container in background.

on how to maintain it. If you can find worms native to your area, that's ideal.

To maintain earthworms, a container as small as a 2-quart (2 l) tub will do (see Figure 8.4), although the bigger the better. Get the lightest, fluffiest soil you can find. Ideally it should be black or dark brown and contain lots of organic matter. Potting soil, good garden soil, or the top layer of forest soil is fine. Before you put the worms in, mix bits of decaying leaves (not oak) with the soil. Maple leaves are good. The leaves should be decayed enough to be in small pieces. The soil should be damp but not sodden. Add more leaf litter, and spray well with a plant sprayer every few days. My worms seem to love used coffee grounds, which I feed them every few days by mixing the grounds slightly into the surface of the soil. You can add other foods as desired instead of or in addition to leaves (see Experiment 10 for suggestions).

Night crawlers kept in a small container will convert much of the soil to castings within a few weeks or less. Smaller worms will too, but not as quickly. The soil will look lumpy and compacted and less fluffy when it is all castings. Before this happens put in fresh soil and leaves. If any worms become injured by students, take them out. A dead and rotting worm smells awful.

Worms at School

Getting Ready

There are so many different things to do with worms—the simplest require no preparation other than clearing off a table or a desk for everyone to gather around.

Observations and Activities

The most fun thing I've done with worms has been taking about ten night crawlers to a kindergarten class. We put two tables together so that about thirteen students can gather around at once, and I take half the class at a time. I put three or four night crawlers on the table and ask the students to watch, not to touch. Night crawlers will leave castings on the table, so be sure to discuss them in advance. Tell the students that the worms will be leaving little piles of "processed soil" behind them, which is one reason the worms are so good for gardens. Tell them the processed soil is called castings and contains nutrients that plants need. You may want to explain that some people think worms are "icky" because they just don't know much about them.

Fig. 8.5 Michael gives a night crawler the one-finger touch to start.

Here is a list of questions that will help the students observe the worms carefully, many of which are from Molly McLaughlin's excellent book *Earthworms, Dirt, and Rotten Leaves.* How does the worm move? Encourage the students to notice how it stretches its body out, then pulls it up. It gets long and thin, then short and fat. Does the worm have legs? Eyes? Ears? Nose? Mouth? Stripes or rings? Is there a difference between head and tail (point out the clitellum toward the head end)? Is there a difference between top and bottom (it may be darker on top or may not)? How does its skin feel? Do you think the worm has a skeleton? What does the worm do when it comes to a block? To a pile of soil? What does it do when you hold it? Who has ever found a worm outside? Where was it? What shape is a worm? Why is this a good shape for it? Does it need legs or sense organs? Why not? Do we? Why?

Eventually, or maybe immediately, one or a few students will ask to hold a worm.

Fig. 8.6 Jolyn's not squeamish in the least!

Those who are wary may be comfortable just touching at first (see Figure 8.5). My experience is that once one or two children have held the worms, almost everyone feels encouraged to try. They may soon be wrapping worms around their ears and draping worms over their noses with glee (see Figure 8.6)!

Experiments

Remember that the hypotheses I give are just examples. Your hypotheses will be the predictions made by the class or a particular student. Your result for each experiment will be a statement of how your animals reacted to your experimental setup. Your conclusion is a statement of whether your prediction was confirmed or not. For each experiment, adding replicates increases your confidence in the validity of your conclusion, but they may be omitted if tedious for young children.

Experiments 1 to 4 explore the types of living conditions that worms prefer.

Experiment 1

Question:
Do worms prefer damp or dry conditions?

Hypothesis:
We think worms prefer dampness.

Methods:
I use a 9-by-13-inch (22.9 x 33 cm) baking pan, but any sort of tray will do. Put a damp paper towel in one end of the tray and a dry paper towel in the other end, so that they almost meet in the middle. Put the worm into the middle. Wait until the worm has settled down on one side or the other and record the results. Repeat several times with the same worm or other worms. You can also do this experiment by setting the worms down between a pile of damp soil and a pile of dry soil in a box. You can record your results on a table (see Figure 8.7 for an example). Figure 8.8 provides a blank table on which to record your results. You can also record your results as a histogram in Figure 8.9. Figure 7.4 provides an example of a completed histogram. A final option is to create a line graph (see Figure 8.10 for an example). Figure 8.11 provides a blank graph on which to plot your data.

Result:
Your result is a statement of your worms' choices.

Conclusion:
The prediction is confirmed. In my experience, earthworms definitely prefer dampness. They will burrow under the damp paper towel if there are any blips along the edge of it. Worms breathe

Choice of Conditions Offered	Worms' Choice	Effect of Choice on Worms
Dampness or Dryness (Experiment 1)	Dampness	Keeps skin moist
Warmth or Cold (Experiment 2)	Cool	Keeps skin moist
Darkness or Light (Experiment 3)	Darkness	Keeps skin moist, avoids predators
Loamy Black Soil vs. Sand or Acid Soil or Alkaline Soil (Experiment 5)	Loamy black soil	Easy to tunnel in, not irritating to skin

Figure 8.7 Sample table summarizing results for Experiments 1, 2, 3, and 5.

Choice of Conditions Offered	Worms' Choice	Effect of Choice on Worms
Dampness or Dryness (Experiment 1)		
Warmth or Cold (Experiment 2)		
Darkness or Light (Experiment 3)		
Loamy Black Soil vs. Sand or Acid Soil or Alkaline Soil (Experiment 5)		

Figure 8.8 Blank table for recording results for Experiments 1, 2, 3, and 5.

through their skin, and the skin must stay damp for them to breathe, so they are very vulnerable to dehydration.

Experiment 2

Question:
Do worms prefer warmth or cold?

Hypothesis:
We think worms prefer warmth.

Methods:
I use a rectangular metal cookie sheet, about 16 inches (40.6 cm) long. I put a heating pad set on "high" (cloth cover removed) under one end of the cookie sheet and an ice tray full of ice cubes under the other end. Make the cookie sheet level. Put a single layer of damp paper towels flat on the surface of the cookie sheet so the worms won't try to escape in search of moisture.

You need to allow a half hour or more for the cookie sheet to change temperature before starting.

Have a student place one or more worms in the middle of the cookie sheet. Allow the worms a half hour or more to settle down.

Result:
Your result is a statement of your worms' choices. You can record your results on a table (see Figure 8.7 for an example). Figure 8.8 provides a blank table on which to record your results. You can also record your results as a histogram in Figure 8.9. Figure 7.4 provides an example of a completed histogram. A final option is to create a line graph (see Figure 8.10 for an example). Figure 8.11 provides a blank graph on which to plot your data.

Conclusion:
The prediction is not accepted. In my experience, most worms come to rest toward the cooler end of the cookie sheet.

Experiment 3

Question:
Do worms prefer darkness or light?

Hypothesis:
We think the worms prefer light. (Students are usually anthropomorphic and predict "light" unless they've learned from other chapters to consider what it's like in the places the animal is found naturally.)

Methods:
Any dish big enough for the worm to move around in will do for this experiment. For small worms I use petri dishes, which you can order (see Appendix). Line the dish with a damp paper towel. (If the worms insist on hiding under the paper towel, take it out.) Cover half the dish with black construction paper or anything opaque. If you're using a petri dish, the students can tape the edge of the paper to the midline of the lid. If you're using a kitchen dish, just lay the paper across to cover half of the dish. The paper must be low enough to make that side pretty dark. Put the worm in the middle of the dish. Wait until the worm has stopped investigating and has chosen a spot.

Result:
Your result is a statement of your worms' choices and any relevant observations. You can record your results on a table (see Figure 8.7 for an example). Figure 8.8 provides a blank table on which to record your results. You can also record your results as a histogram in Figure 8.9. Figure 7.4 provides an example of a completed histogram. A final option is to create a line graph (see Figure 8.10 for an example). Figure 8.11 provides a blank graph on which to plot your data.

Conclusion:
In my experience, most worms eventually settle in the darkness.

Help the students relate this to where the worms are found. This preference may be puzzling to them if it occurs to anyone that worms have no eyes. How can a worm tell light from dark? Worms have light-sensitive cells in the skin on their heads. They are not able to form an image as our eyes do but they can detect light, an ability that enables them to stay hidden during the day. They come out of their burrows only at night to get leaf fragments and other edibles; darkness helps protect them from predators. A worm out of its burrow is completely defenseless! It can't bite, can't sting, fly, or jump. Their coloration doesn't disguise them, they're not spiny or tough or hard to eat, and they don't taste bad. Many animals eat worms. Have the students name a few: moles, shrews, raccoons, opossums, toads, frogs and birds. Darkness is earthworms' only protection when out of their burrows.

If you shine a flashlight on a worm halfway out of its burrow at night, it will retreat into the burrow. If you shine a flashlight on a worm crawling across a table, it will recoil from the beam of light, sometimes even when the room is already light.

Try shining the light on the worm's (1) head, (2) midsection, and (3) tail. In which situation does the worm react the most strongly? You might have the students do this as part of a process for showing that worms detect light without eyes because the light-sensitive cells are on their "head."

Experiment 4

Question:
Do worms react equally to light beams of different colors?

Hypothesis:
We think worms dislike all colors of light.

Methods:
Cover a flashlight with red cellophane, then blue cellophane. In a darkened room, illuminate a worm with white light, red light, and then blue light. Does the worm react the same to all?

Result:
Your result is a statement of your worm's reactions.

Conclusion:
In my experience, worms react less or not at all to red light. Worms don't detect red light very well, so a red light would be best for hunting worms at night.

Experiment 5

Question:
What type of soil do worms choose?

Hypothesis:
We think worms prefer loamy black soil.

Methods:
Put two or more different and separate types of soil in a flat container, such as a baking dish. For example, you might have sand in one-half of the dish and loamy, black garden soil in the other. You can also use clay or various mixtures of soil, sand, and clay with or without leaf fragments. You can also alter the soil by mixing a little vinegar with the soil in one end to acidify it. Does the worm prefer the acid soil or the nonacid soil? Make one end alkaline by mixing a little baking soda solution with the soil, and offer that opposite the regular soil. Let the students choose what to compare. Can they offer different combinations and make a ranking of the worms' preference?

Result:
Your result is a statement of your worms' choices. You can record your results on a table (see Figure 8.7 for an example). Figure 8.8 provides a blank table on which to record your results. You can also record your results as a histogram in Figure 8.9. Figure 7.4 provides an example of a completed histogram. A final option is to create a line graph (see Figure 8.10 for an example). Figure 8.11 provides a blank graph on which to plot your data.

Conclusion:
The prediction is confirmed. In my experience, worms avoid sand, clay, acid, and alkaline soil when given loamy black soil as an alternative.

Experiment 6

Question:
If we put a worm in a clear container filled with horizontal layers of different soil types, in which layer will the most tunnels appear?

Hypothesis:
We think the worm will tunnel most in black loamy soil.

Methods:
You need a clear, deep, flattened container for this, so that the worms will be forced to make their tunnels next to the wall where they will be visible. The homemade container in Figure 8.2 is made of two sheets of Plexiglas or glass only $^3/_8$ inch (1 cm) apart, mounted in a wood frame. Fill the container to about 8 inches (20.3 cm) from the top

Number of Worms

Fig. 8.9 Blank histogram for Experiments 1, 2, 3, and 5.

- *For Experiment 1, label two histogram bars "Choosing Dampness" and "Choosing Dryness" as shown in the completed histogram. Number the vertical axis to match your sample size.*
- *For Experiment 2, label the three histogram bars "Choosing Warmth," "Choosing Cold," and "In the Middle."*
- *For Experiment 3, label two histogram bars "Choosing Darkness" and "Choosing Light."*
- *For Experiment 5, label one histogram bar "Choosing Loamy Black Soil." Label the other(s) "Choosing Sand" or "Choosing Acid Soil" or "Choosing Alkaline Soil."*

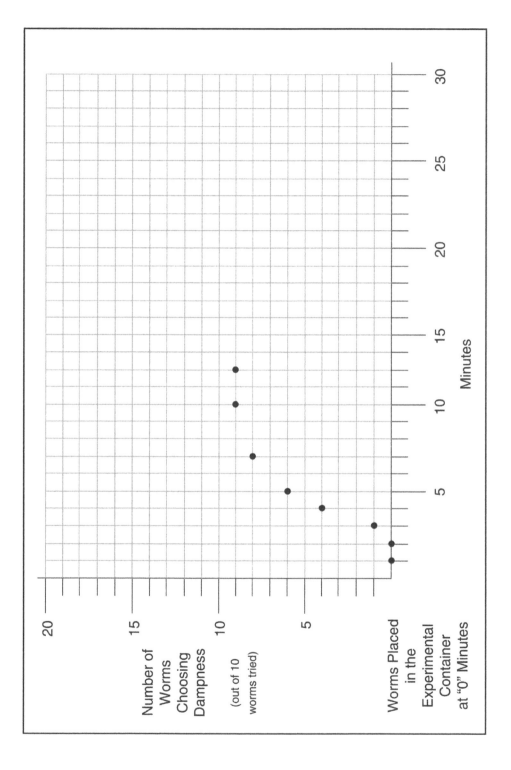

Figure 8.10 Sample graph showing the number of worms choosing dampness over a period of time (Experiment 1).

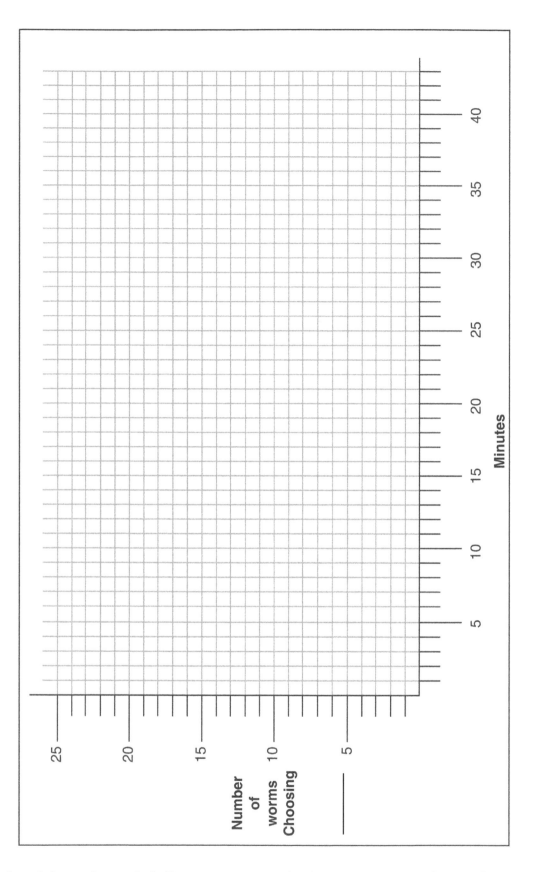

Figure 8.11 Blank graph for recording results for Experiments 1, 2, 3, and 5. Change the numbers on the vertical axis to match your sample size for each experiment, if needed.

- *For Experiment 1, put "Dampness" in the blank on the vertical axis label.*
- *For Experiment 2, put "Cool" in the blank on the vertical axis label.*
- *For Experiment 3, put "Darkness" in the blank on the vertical axis label.*
- *For Experiment 5, put "Loamy Black Soil" in the blank on the vertical axis label.*

THE EFFECT OF EARTHWORMS ON THE SOIL

Effect	Increased by Worms	Decreased by Worms
Number of Tunnels in the Soil (Experiment 6)		
Mixing of Layers of Soil (Experiment 6)		
Volume of Soil (Experiment 7)		
Health of Plants (Experiment 8)		

Figure 8.12   Blank table for recording results for Experiments 6, 7, and 8.

with each type of soil you want to test, in turn. I've used repeating layers of loamy black soil, sand, and a soil-clay mixture. Put a healthy worm on top of the soil. Watch the worm for a few minutes, then put the cage in a dark place. After twenty-four hours, measure the total length of the tunnels it made. (You can have the students calculate the amount of tunneling that would occur in one week, one month, or one year at that rate.) In which type of soil do they tunnel most?

Result:

Your result is a statement of which layer had the most tunnels and any other relevant observations. You can record your results on the table in Figure 8.12.

Conclusion:

In my experience, earthworms tunnel most in black loamy soil that they can easily push through. Is there evidence that the worm's tunneling mixes the layers? Sometimes you can see that worms drag soil with them when they move from one soil type to another.

Experiment 7

Question:

Can worms convert food into soil?

Hypothesis:

We think worms can make soil if we give them what they need to eat.

Methods (from Dorothy Hogner's book, *Earthworms*):

Put a few rocks or pieces of broken flowerpot in the bottom of a jar. Fill the jar three-fourths full of damp sand or sandy soil. Add two to twelve worms. On top of the sand sprinkle food for the worms: 1 teaspoon (5 ml) coffee grounds, $^1/_4$ teaspoon (1.25 ml) brown sugar, cabbage leaves, carrot tops, celery leaves, or maple leaves. Wrap the jar with dark paper that you can remove to observe the worms or leave the jar in a dark room. If you use a red light to observe, you can see the worms moving in the burrows. Feed the worms and moisten the soil regularly for a month or two, then check your results.

To have an experimental control, you would need to set up a second jar simultaneously that is identi-

cal to the first one, but leave out either the food or the worms. This would provide evidence that the soil is a result of the worm and the food.

Result:

Your result is a statement of your findings after a month or two. You can record your results on the table in Figure 8.12.

Conclusion:

The prediction is confirmed. Earthworms will convert the food into a thin layer of soil on top of the sand.

Experiment 8

Question:

Do earthworms in the soil improve the health of plants?

Hypothesis:

We think earthworms in the soil won't affect the plants.

Methods:

Get two plants of the same type and same size, equally healthy, and two flowerpots. Plant the two plants in separate pots in good quality garden soil or potting soil. Add two to four worms to one pot only (more for large plants and pots). Compare the condition and size of the two plants after a month or two.

Result:

Your result is a statement of your observations of the two plants. You can record your results on the table in Figure 8.12.

Conclusion:

The prediction is not confirmed. After a month or two, one plant should be taller and bushier than the other. Earthworms in the soil improve the health of plants by (1) tunneling and thereby aerating and loosening the soil so water can penetrate, and (2) improving the texture and nutrient content of the soil with their castings.

Experiment 9

Question:

Can worms find their food in a maze?

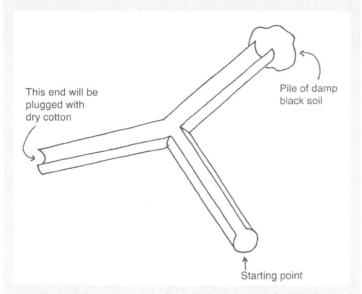

This end will be plugged with dry cotton

Pile of damp black soil

Starting point

Fig. 8.13 A Y-shaped maze made from three paper-towel rolls. Line the trough with aluminum foil and cover the open top with plastic wrap.

Hypothesis:

If we offer worms a forked path, only one fork of which leads to soil, we think they can find the soil.

Methods:

You need a clear Y-shaped tube big enough for a worm to move through. You can buy one, but I made my own tube or maze by using two or three paper towel rolls, plastic wrap, foil, and tape. Cut a lengthwise strip out of each roll so that you could watch a worm crawling through it. You'll have high-walled troughs now instead of rolls. Then trim one end of each so that you can fit them together to make a Y-shaped trough (see Figure 8.13). You may want to shorten them somewhat. Tape them together, then line the trough with foil so it will be smooth inside. Finally, cover the top of the trough with plastic wrap, so you can see in, but the worms can't get out. You'll put the worms in at the base of the Y, so the two tips will be the exits. Plug one exit with cotton balls. Put the other exit hole into a pile of moist savory soil. Now put a worm into the entry hole and block the hole to keep it in. Watch where it goes.

I've used night crawlers with this experiment, but other worms will do.

Result:

Your result is a statement of your observations of your worm's movements.

Conclusion:

The prediction is confirmed, usually. Some worms will take off right away and go right to the soil. Some will just sit there and refuse to go anywhere, not responding to prodding, light, or even tilting the tunnel. Replace them.

The most fun are those who take off and make the wrong choice. My son was worked into an absolute frenzy one day with a pair of night crawlers in the maze. "Mr. Worm" had made the wrong choice and was loitering in the wrong trough, unable to "decide" what to do.

So we released "Ms. Worm" into the entry hole. She immediately joined Mr. Worm. They turned around and returned to the juncture of the three tunnels lying side by side. They then probed the intersection with their heads for several minutes, lying side by side, as though trying to "decide" which way to go. Alan was frantic. Finally Mr. and Ms. Worm took off down the "right" tunnel, side by side, and reached the tasty soil. Alan was jubilant. He couldn't refrain from probing the soil with his fat little fingers to congratulate the worms on their victory.

An individual worm can learn after fifty to a hundred trials to choose the correct tunnel.

Experiment 10

Question:

Do earthworms prefer some foods over others?

Hypothesis:

It depends on what is offered.

Methods:

Place small amounts of different foods on top of the soil, such as cornmeal, green leafy vegetables, grass clippings, potato peelings, coffee grounds, and small bits of other fruits and vegetables. Wait twenty-four hours or longer and note which ones have been taken.

Result:

Your result is a statement of your worms' preferences. You can record your results on the table in Figure 8.14.

Conclusion:
 It depends on the choices offered. Worms will eat almost any fruit or vegetable in small pieces and all of the things suggested here. Offerings like cornmeal or coffee grounds need to be spread out or worked into the soil a bit.

	Not Accepted	Accepted	Not Sure
Cornmeal			
Lettuce Bits			
Bits of Celery Greens			
Grass Clippings			
Bits of Potato Peel			
Bits of Apple			
Bits of Cucumber			

Figure 8.14 Blank table for recording results for Experiment 10.

Chapter 9
Chirping Crickets

Introduction

Cricket studies range from the very simple to the rather complex. This chapter describes several activities and experiments appropriate for children as young as four or five and others that will work better with somewhat older students. When I first began to learn about crickets, I was surprised at the intricacy of their aggressive behaviors. As I write this a pair of males sit on my desk head to head, alternating aggressive chirps. The children's interest in the chirps is, by itself, enough reason to keep a terrarium of crickets in your classroom. But there's more to it than that.

After you've done the experiments in this chapter and in Chapters 7, 8, and 9, refer to the table in Figure 1 in the Postscript, which provides a comparison of the results for the experiments in these chapters. Figure 2 in the Postscript is a blank table that can be filled out with the results of your experiments, for comparison.

Materials

1. Male and female crickets. Crickets are easy to catch, also easy to buy locally as fish bait (details provided later in the chapter).

2. One to several terraria, depending on how many activities and experiments you plan to do. Size of the terraria is flexible. I use some as small as $5^1/_2$ by 3 inches (14 x 7.6 cm) and others five times that size.

3. Sand for the floor of the terrarium.

4. Plastic dish, such as trimmed cottage cheese or yogurt containers, to hold wet sand for egg laying.

5. Wet bread and lettuce for cricket food.

6. Bottle of correction fluid for marking crickets.

7. Various materials like toilet paper rolls, toothpick boxes, or matchboxes to make houses for the crickets. One experiment requires a piece of clear flexible plastic like the transparencies used for overhead projectors.

Background Information

Crickets have recently become a favorite activity around my house. My four-year-old Alan sits by the cricket terrarium for a half hour at a time handling them, using a pencil to make tunnels for them in the wet sand, and moving their cardboard houses around. He and his sister have both demanded separate enclosures for their personal crickets. Sadie wants females only, please. They boast to one another, "My cricket is laying eggs." Alan watches his chirp, court, and lay eggs while he eats his breakfast, filling me in on their every move.

Crickets are often used in college laboratories to demonstrate aggression and territoriality in insects, which is why I decided to investigate crickets as experimental subjects for children. I was well acquainted with crickets as food for praying mantises but had never kept them long enough to watch their behavior. Now after keeping two species of crickets on my kitchen table for a couple of months and sharing them with my own children, a class of kindergartners, and two classes of older students, I see that their various aggressive behaviors are not the features that most attract children to them. The aggression—other than aggressive chirping—doesn't occur frequently enough to catch the children's attention. I learned what does interest them by watching the children watch the crickets.

Most of the simple experiments I describe here didn't occur to me until I saw the crickets through the children's eyes. Many of the questions they asked were anthropomorphic or simplistic ("Do crickets make friends with other animals?"), but many were not. The many that were not led to some interesting discoveries for us all. I give much more information about the crickets themselves in the following sections as it relates to particular classroom setups for observation and to particular experiments.

How to Get and Keep Crickets

Most big pet stores sell crickets as food for pets, but crickets sold as fish bait can be cheaper. Look under "fisherman's supplies" or "fishing tackle" or "fishing bait" in your town's Yellow Pages, or on the computer. You can also order crickets from biological supply companies (see Appendix), but that's probably more expensive than buying them locally.

Crickets are easy to catch too, especially if you set up an area in advance that they'll like. Keep a pile of slightly damp grass clippings near a brushy area or on the edge of your yard. When I look through my neighbor's grass clipping pile, I can find ten crickets in ten or fifteen minutes every time. The crickets are quick—there's definitely a trick to catching them.

I find that if I pull back a big hunk of grass clippings and just wait for a minute, a cricket or two will come out into the open space. I have a plastic peanut butter jar ready and slap it down over the cricket. Cover the opening with your hand while you turn the jar right side up. Have a second container with you as a holding jar. You'll miss the first few but then you'll get the hang of it. I find this technique much easier than trying to catch them by hand.

I also find crickets under the black plastic sheet that covers my compost pile, under my outdoor garbage can, in my children's damp sandbox under the toys, under the edge of the grass that abuts my wooden garden border, and under the leafy plants in the flower garden. By far the most consistently successful place for cricket collecting for me, however, is the pile of damp grass clippings. This particular pile is on a slope above a ditch that sometimes has water in it; the ditch may contribute to its success.

All the crickets I've worked with have been field crickets or house crickets, which look similar and are very closely related. Much of what I've written here will apply to any crickets. The behavior of different species, however, is not identical, and you may find differences. I've never been bitten by a cricket, although some grasshoppers will bite.

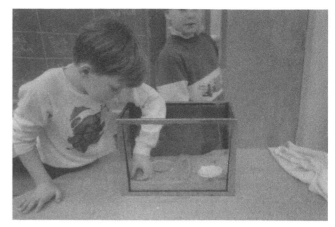

Fig. 9.1 *Malcolm reaches for a subject in the terrarium.*

Maintaining Crickets

I keep crickets in a terrarium with about 2 inches (5 cm) of sand on the floor (see Figure 9.1). Most of the books I've looked at on keeping crickets suggest sand, although it isn't essential. They do like to dig sometimes, and children may want to make tunnels for them in damp sand. If you want them to lay eggs you'll have to have at least a small area of damp sand. Otherwise dry sand or soil is okay. Adults may eat the hatchlings, so if you know you have eggs in an area, you many want to cover that area with a wire mesh cage or put the adults somewhere else. I've read that they're less likely to cannibalize if you give them plenty of other protein, like dog food, but my crickets won't eat dog food.

To provide water, you can simply spray the glass with water droplets once a day. Most insects drink from dewdrops in their natural habitat. Or you can put a small dish of shallow water ($^1/_8$ inch or .3 cm maximum depth) in the terrarium. Use something with short sides like a cap from a medicine bottle. Crickets will eat wet bread, moistened Cheerios and other cereal, lettuce, carrots, and probably many other foods. Those are the items I feed mine. I keep food available at all times. You can remove it for a day or two to conduct an experiment, but you don't have to.

Field Hunt

If you want to take students to look for crickets, any brushy area or area with leaves instead of lawn will do. Look under the leaves and under logs, in crevices in logs, under boards, under anything. You may find spiders and snakes in the same places, but don't let this stop you from looking. You probably won't. Still you should caution students about not picking up spiders and snakes, and you may want to turn logs and boards for them. It's never a good idea to stick fingers into dark places you can't see into. Students enjoy going after the crickets with their hands as well as with jars.

Crickets at School

Getting Ready

The minimum you need to begin is a terrarium, sand for the terrarium floor, crickets, and bread to feed them.

Observations and Activities

Before you give the students any information about the crickets, it's a good idea to let them watch the crickets at their leisure for a few days. Start with one male. Ask the students some questions to help them notice his behavior. What is he doing? He wanders around at first. Does he chirp? He may a little. Does he take cover? He will after he looks around, if cover is available. Does he show any courtship behavior (see description in the following paragraph)? Not if he is alone. Does he eat and drink?

After a day or two add a female. Does the male's behavior change? Does he chirp now? He probably will begin to court her. Courtship behavior consists of, for one thing, chirping softly by gently rubbing his wings together. (The wings have files on them that produce the chirping sound when rubbed together.) His wings will still be almost flat against his back during courtship chirping, not raised as they are in aggressive chirping. I watched courting males of the species rock their bodies back and forth gently while chirping and claw at the sand gently with all legs continually. A courting male will eventually back up to the individual he is courting, to entice her onto his back. If she does get on his back, he will curl the tip of his abdomen up and transfer a sperm packet to an opening in the tip of her abdomen. The sperm packet is about half the size of a tomato seed and gray. Sometimes, if they are interrupted, the sperm packet will get stuck on one of them and be carried around.

Does the female accept his advances? Usually not. If he approaches her from behind, she may kick him in the face! Females you buy from a bait store have already mated and are full of eggs because they've had nowhere to lay them. In my experience they won't mate again until they find a place to deposit the eggs they already have. A female must have damp sand to lay her eggs. If she does have damp sand, she will almost certainly lay eggs in it unless she has been isolated from males previously for a good while. Eggs come out singly. She will continue to lay eggs intermittently and often throughout her lifetime.

After another day or two add an additional male. I like to mark my males with model airplane paint so that I can tell them apart. To mark one, I put the male in the bottom of a very small juice glass, with a bottom diameter of no more than $1^1/_4$ inch (3.2 cm) or so, so that he can't move around very much. Then I follow him around with the paintbrush until I can get a tiny spot on his thorax. It doesn't bother him—but don't get it on his abdomen because they breathe through pores in the abdomen.

You can also mark a male by clipping one of his antennae. If you get them from the bait store, some will probably already have broken antennae.

At least one of the two males will probably behave aggressively toward the other, but usually not right away. They may face off and chirp aggressively, or one may stalk the other while chirping, causing the other to retreat. The students may notice that an aggressive chirp is often louder than a courtship chirp. You can also see that the wings move higher to make the aggressive chirp. Males often court other males. A male is stimulated to begin courtship when he encounters another cricket and the other cricket does not respond to him aggressively, whether that cricket is male or female. In nature, a nonresponsive cricket is probably usually a female, so their behavior is usually appropriate. Write down all questions the students have during this initial period of observation.

After a couple of days of watching the terrarium with two males and a female in it, take an hour or so to gather the children around and talk to them about the crickets. Show them all the body parts (see the following list and Figure 9.2).

1. Big legs for jumping to escape from predators, like their relatives the grasshoppers.

2. Wings, which when rubbed together make a chirping sound.

3. Long antennae. All crickets feel one another with their antennae any time they meet. Males may lash one another with their antennae during low-level aggressive encounters. You can induce aggression in a male by lashing him with a fake antenna.

4. Cerci, which are the two prongs that project from the back of the abdomen. Stimulation of the

cerci plays a part in courtship. It signals him that another cricket is behind him. He may back up, trying to back under her, to get her on his back. If I touch a male's cerci with my finger while he's courting, he kicks my finger repeatedly but doesn't run away! (The cerci are not genitals or reproductive organs, just feelers.)

5. Ovipositor (egg depositor). The female has a long sword-shaped projection from the back of her

abdomen, between her cerci. She jabs it down into the wet sand and lays her eggs from the end of it. You can sometimes see the eggs come out when she pokes it between the sand and the side of the terrarium. The eggs are tiny and are laid singly. The male crickets will feel disturbed and won't court or chirp or show any aggressive behavior while the children are all gathered around the terrarium, but the females will still lay eggs.

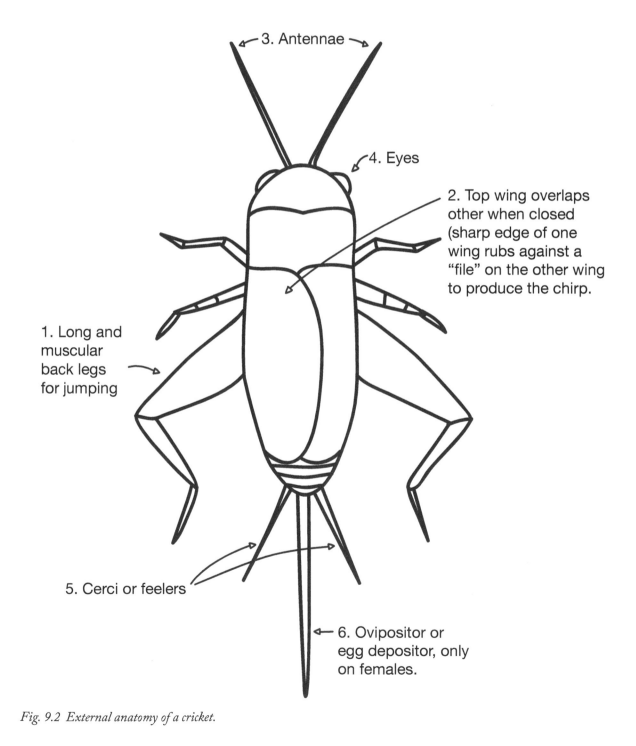

Fig. 9.2 External anatomy of a cricket.

I've found that what young children like best is holding the crickets and letting the crickets run up their arms (see Figure 9.3). A class of kindergartners and I had been watching the crickets for about a half hour as a group when a little boy asked if he could hold one. I told him to go ahead, so he reached into the terrarium and caught one. Pretty soon half the class was asking to hold one. The first few bold souls simply held the crickets stiffly in their hands and then dropped them back into the terrarium. But a little red-head named Brian opened his fist and let his cricket wander up his arm, across his shoulders, across his back, and on and on. He inspired the rest of them. Soon everyone was transferring or receiving crickets, amid squeals of eagerness. It was one of the most exhilarating experiences I've had with young children and creatures. They were enraptured.

Fig. 9.3 Alan grabs a cricket by the front legs, showing off its cerci.

As the time came to put away the crickets and get ready to go home, the students gathered in a circle to talk. Little red-haired Brian summed up the cricket experience aptly: "Something interesting happened today!"

Another thing children seem to enjoy particularly is being able to identify and name the individual males. I put a card on the wall by the terrarium showing several roughly drawn cricket shapes. On each shape was the pattern of paint spots for an individual cricket, labeled with the name of the particular cricket with that paint pattern.

The students watch them and say, "Oh, Larry is chirping at Thor again," or "Fred stays in that little house by himself almost all the time."

One of the first things children will notice about crickets is their tendency to burrow. If there is any wet sand in the terrarium, some of the crickets will probably dig into it. You can encourage burrowing by keeping a pile of loose wet sand in the corner of the terrarium. They'll be able to burrow into the pile without hitting the bottom of the terrarium.

Crickets like already-made burrows too. Poke a pencil horizontally through a pile of damp sand in the terrarium. With the pencil in place, pack the sand firmly over it. Slowly withdraw the pencil, leaving a tunnel with two openings. The students can do this, but make sure they pack it well or the tunnel will collapse and smother the crickets.

If you have areas of wet sand as well as dry sand, the students will notice that the crickets prefer to rest on the wet sand. They'll also notice that the crickets take refuge whenever it (a matchbox, toilet paper roll, etc.) is available. Toilet paper rolls are real favorites.

The simple experiments I've done with very young children are derived from their own simple observations and questions, which is why I write down every question. Letting their questions be your guide is a good way to avoid getting too complex for their particular age.

Below are descriptions of the simple experiments I've done with very young children; following are some that are a bit more complex.

Experiments

In each experiment, unless otherwise specified, the terraria should all have a floor of damp sand or damp soil with wet bread or other food available. Provide a water dish or spray the inside walls with water droplets once a day.

Remember that the hypotheses I give are just examples. Your hypotheses will be the predictions made by the class or a particular student. Your result for each experiment will be a statement of how your animals reacted to your experimental setup. Your conclusion is a statement of whether your prediction was confirmed or not.

For each experiment, adding replicates increases your confidence in the validity of your conclusion, but they may be omitted if tedious for young children.

Experiment 1

Question:

Do crickets prefer houses with two doors or one door?

Choice Offered to Crickets	Crickets' Choice	Effect of Choice on Crickets
1-Door House vs. 2-Door House (Experiment 1)	2-door house	Have escape route
Empty House vs. House with Company (Experiment 2)	Depend on crickets' recent experiences	Solitude reduces conflict between males
Damp Substrate vs. Dry Substrate (Experiment 3)	Damp	Reduces dehydration
Dark Tube vs. Clear Tube (Experiment 4)	Dark	Reduces dehydration
Dark End of Terrarium vs. Light End of Terrarium (Experiment 5)	Light end (always?)	Darkness reduces dehydration and predation
Damp Sand for Tunneling vs. Dry Sand for Tunneling (Experiment 6)	Damp sand for tunneling	Damp sand holds shape better
Moss vs. Damp Grass Clippings (Experiment 7)	Damp grass clippings	Reduces dehydration; easier to burrow into

Figure 9.4 Sample table summarizing the choices of the crickets in Experiments 1 to 7.

Choice Offered to Crickets	Crickets' Choice	Effect of Choice on Crickets
1-Door House vs. 2-Door House (Experiment 1)		
Empty House vs. House with Company (Experiment 2)		
Damp Substrate vs. Dry Substrate (Experiment 3)		
Dark Tube vs. Clear Tube (Experiment 4)		
Dark End of Terrarium vs. Light End of Terrarium (Experiment 5)		
Damp Sand for Tunneling vs. Dry Sand for Tunneling (Experiment 6)		
Moss vs. Damp Grass Clippings (Experiment 7)		

Figure 9.5 Blank table for recording results for Experiments 1 to 7.

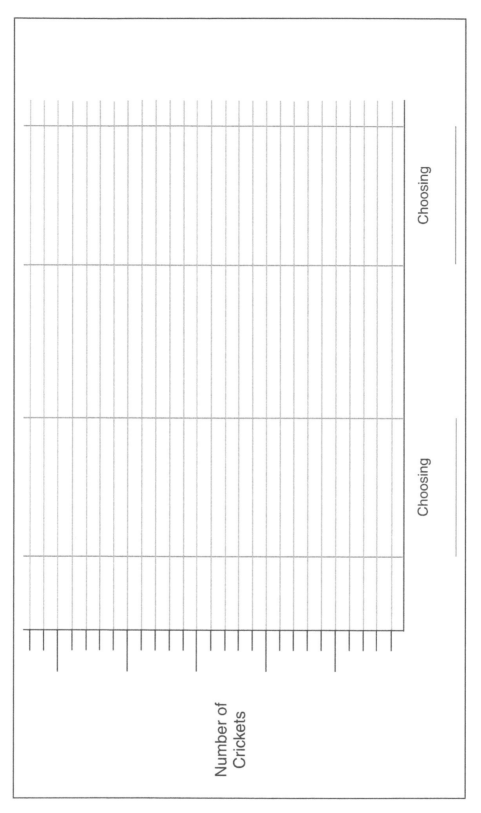

Figure 9.6 Blank graph for recording results for Experiments 1 to 7. (See Figure 7.4 for an example of a completed histogram.)

- *For Experiment 1, fill in blanks on horizontal axis with "1-Door House" and "2-Door House."*
- *For Experiment 2, fill in blanks with "Empty House" and "House with Company."*
- *For Experiment 3, fill in blanks with "Damp Substrate" and "Dry Substrate."*
- *For Experiment 4, fill in blanks with "Dark Tube" and "Clear Tube."*
- *For Experiment 5, fill in blanks with "Dark End of Terrarium" and "Light End of Terrarium."*
- *For Experiment 6, fill in blanks with "Damp Sand" and "Dry Sand."*
- *For Experiment 7, fill in blanks with "Grass Clippings" and "Moss" or "Damp Sand."*

Hypothesis:

We think crickets will like two-door houses and one-door houses the same.

Methods:

Make several houses out of toothpick boxes or matchboxes or toilet paper rolls cut to about 2 inches (5 cm) in length. Put a corresponding number of crickets into the terrarium. Have only one opening in most of the houses, perhaps all but one house. In the remaining house make two openings on opposite ends. Which type of house do the crickets prefer?

Result:

Your result is a statement of your crickets' choices. You can record your outcome on a table (see Figure 9.4). Figure 9.5 provides a blank table for your results. You can also make a histogram with Figure 9.6, using Figure 7.4 as an example of a completed histogram.

Conclusion:

The prediction is not confirmed. My experience is that the crickets usually prefer two doors. The last time I did this experiment with field crickets, all six males stayed in the two-door house, ignoring the one-door houses. Apparently their preference for two doors outweighed their tendency to be solitary and aggressive toward other males. Two doors are obviously safer in terms of predators. An insect-eating shrew comes in the front door, the cricket goes out the back. Many animals are uncomfortable when they have no potential escape route.

Experiment 2

Question:

Will crickets choose houses with other crickets or will they house themselves singly?

Hypothesis:

We think crickets will choose to have company.

Methods:

Make several houses (see Experiment 1). Have two openings in each house. Put as many crickets in the terrarium as you have houses. Do they choose separate houses or all inhabit the same one, or do they inhabit none?

Result:

Your result is a statement of your crickets' choices and any relevant observations. You can record your outcome on a table (see Figure 9.4). Figure 9.5 provides a blank table for your results. You can also make a histogram with Figure 9.6, using Figure 7.4 as an example of a completed histogram.

Conclusion:

It's hard to predict the outcome of this one, which makes it interesting. Crickets do prefer a refuge, and in nature they live singly and rather widely spaced, which is one reason the males chirp (to help females find them). But being crowded for long periods in captivity seems to affect their tendency to be solitary. In this situation, they probably will move into the houses separately, but they may not. You can try isolating the crickets for several days before the experiment. Does this affect the outcome?

Experiment 3

Question:

Do crickets prefer a damp or a dry substrate?

Hypothesis:

We think crickets will like damp sand and dry sand equally.

Methods:

Pour water over the sand in one end of the terrarium. Leave the other end dry. Count the number of crickets in each location after the class has been absent from the room for a while. Activity outside the terrarium causes the crickets to mill around.

Result:

Your result is a statement of your crickets' choices and any other relevant observations. You can record your outcome on a table (see Figure 9.4). Figure 9.5 provides a blank table for your results. You can also make a histogram with Figure 9.6, using Figure 7.4 as an example of a completed histogram. Finally, you can plot your count on a graph (see Figure 9.7). Figure 9.8 provides a blank graph for your results.

Conclusion:

The prediction is not confirmed. Crickets have a definite preference for damp sand over dry sand. A damp environment helps them avoid dehydration.

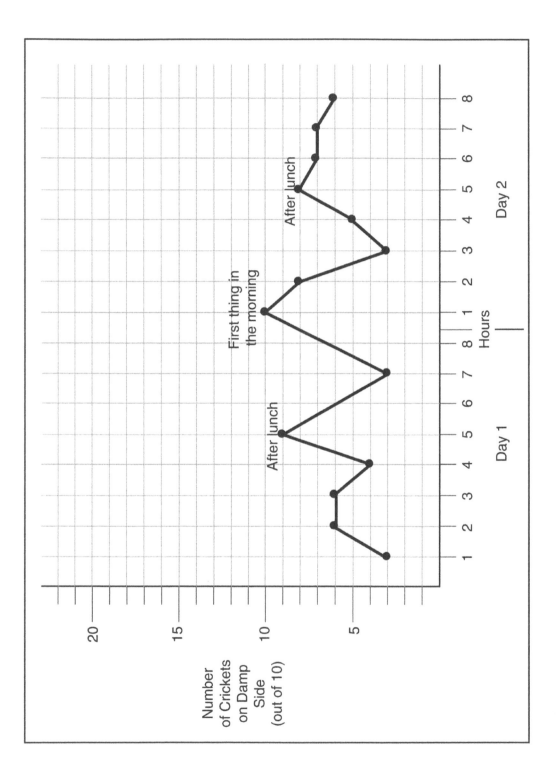

Figure 9.7 Number of crickets choosing dampness in Experiment 3. The peaks are after the crickets have been undisturbed for a while.

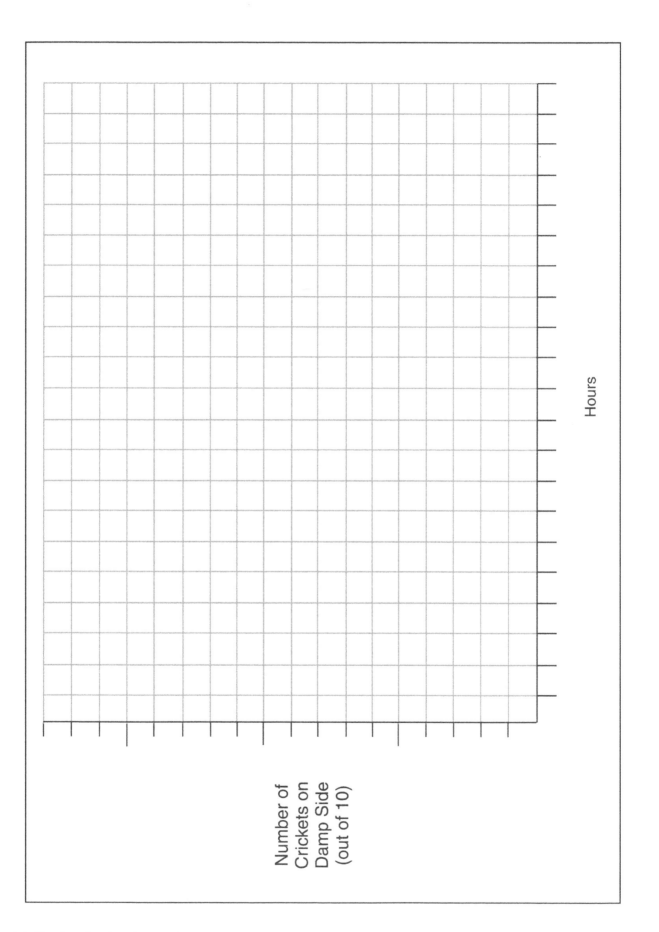

Figure 9.8 Number of crickets choosing dampness in Experiment 3. Fill in the numbers on the horizontal axis according to how long you watched the crickets.

Experiment 4

Question:

Do crickets stay in houses because they are attracted to darkness or because they like being enclosed?

Hypothesis:

We think crickets are attracted to darkness.

Methods:

Animals can be attracted to the feel of something around them, an urge that is called thigmotaxic. Roly-polies, for example, are thigmotaxic (see Chapter 7). They seek out crevices or other bodies to make contact with. When a cricket is inside a toilet paper roll, you can see that its antennae are making contact with the cardboard over and around it, so it is receiving physical signals that something is surrounding its body. Do crickets stay in the roll because of the darkness or because of the physical contact?

To address this question you need to offer something that provides contact but not darkness. I made a clear tube open on both ends. As a control I offered at the same time a black tube open on both ends. You can make a clear tube out of a sheet of clear or translucent plastic—maybe a plastic sheet protector. A dark tube can be made from black construction paper. As another option I've used two *Drosophila* culture vials (see Chapter 16) that have had the closed ends sawed off. Leave one clear, and cover the other one with duct tape. Lay the dark tube and the light tube horizontally on the terrarium floor. Put some sand in the floor of each tube to anchor it.

If the instinct to be surrounded by something is responsible for the crickets' behavior, then they should like the clear tube just as well as the dark tube. Wait several hours or until most of the crickets have chosen one tube or the other.

Result:

Your result is a statement of your crickets' choices and any relevant observations. You can record your outcome on a table (see Figure 9.4). Figure 9.5 provides a blank table for your results. You can also make a histogram with Figure 9.6, using Figure 7.4 as an example of a completed histogram.

Conclusion:

The prediction was confirmed here. In my experience, most crickets prefer the dark tube. So, we may extrapolate and assume that crickets stay in their houses a lot because, at least in part, they are attracted to darkness. They may also be thigmotaxic.

The answer is not really that simple! (See the following experiment.)

Experiment 5

Question:

Are crickets attracted to light?

Hypothesis:

We think crickets are not attracted to light, since they chose the dark tubes in Experiment 4.

Methods:

Put a lamp beside one end of the terrarium so that it shines down into that end only. Wait an hour or so and observe your crickets' positions.

Result:

Your result is a statement of your observations. You can record your results on a table (see Figure 9.4). Figure 9.5 provides a blank table for your results. You can also make a histogram with Figure 9.6, using Figure 7.4 as an example of a completed histogram.

Conclusion:

Within a few hours the crickets will all congregate at the illuminated end of the terrarium, obviously in response to the lamplight. I noticed this by accident when keeping a terrarium next to a lamp. I can think of only one explanation: The crickets are attracted to the warmth of the lamp instead of the light itself. You could test this by using alternately an incandescent bulb that gives off a lot of warmth and a fluorescent bulb that gives off little warmth. Or put a hot water bottle and a cold water bottle in the terrarium.

If this doesn't explain it, then there is evidently something else involved in their attraction to tubes and tunnels. Maybe your students will think of some other possibility.

Experiment 6

Question:

Will crickets make more tunnels in dry sand or in wet sand?

Hypothesis:

Crickets will make tunnels in either.

Methods:

Wet the sand in part of the terrarium. Give the crickets twenty-four hours or so to make tunnels.

Result:

Your result is a statement of your crickets' tunneling activities in each end of the terrarium. You can record your results on a table (see Figure 9.4). Figure 9.5 provides a blank table for your results. You can also make a histogram with Figure 9.6, using Figure 7.4 as an example of a completed histogram.

Conclusion:

In my experience, crickets will tunnel in the wet sand only. Is this because it sticks together better?

Experiment 7

Question:

Do crickets prefer moss, damp grass clippings, or damp sand?

Hypothesis:

We think they'll like moss because it's soft.

Methods:

Cover one-third of the terrarium floor with moss, one-third with damp grass clippings, and one-third with damp sand. Wait twenty-four hours. Where do the crickets hang out?

Result:

Your result is a statement of where most of the crickets are located after twenty-four hours. You can record your results on a table (see Figure 9.4). Figure 9.5 provides a blank table for your results. You can also make a histogram with Figure 9.6, using Figure 7.4 as an example of a completed histogram.

Conclusion:

In my experience crickets like substrates that are damp and easiest to burrow into, which here means the grass clippings.

Experiment 8

Question:

Do crickets prefer food A or food B?

Hypothesis:

Crickets prefer food A (or whatever the students predict).

Methods:

Offer the crickets choices among these foods: carrot peelings, lettuce, dry dog food, wet dog food, apple peelings, wet bread, dry bread, or anything else the students think of. The above items are those I've seen them eat or read that they will eat. Your result is a statement of your crickets' preferences. Crickets like food B better than food A, and so forth. You can record your results on the histogram in Figure 9.9. Figure 7.4 provides an example of a completed histogram.

Conclusion:

Maybe the students can make some generalizations; for example, crickets like vegetables better than meat, or they like fresh vegetables better than rotting vegetables, or they like thinly sliced vegetables better than thick or raw or cooked, and so forth.

Male Aggression and Territoriality

The aggressive behavior of male crickets is complex and fascinating. It differs in one aspect from the aggressive and territorial behavior of many other animals. Often when a male animal (other than a cricket) establishes dominance over other males, that dominance is based on his superior strength or size or daring or fierceness. He is consistently able to intimidate all of his rivals, which is what is meant by being "dominant." Other males know him and once they've been defeated by him, they generally don't challenge him again. He may be displaced eventually by a newcomer or by an animal that has just reached maturity.

A dominant male cricket is able to intimidate all other males. But the dominance of a male cricket has nothing to do with his size or strength. None of the other crickets recognize him or avoid fights with him. His dominance is not based on any innate quality that makes him a superior fighter. Rather, it is based on what has happened to him in the recent past. Certain transient events in the life of a cricket predispose him

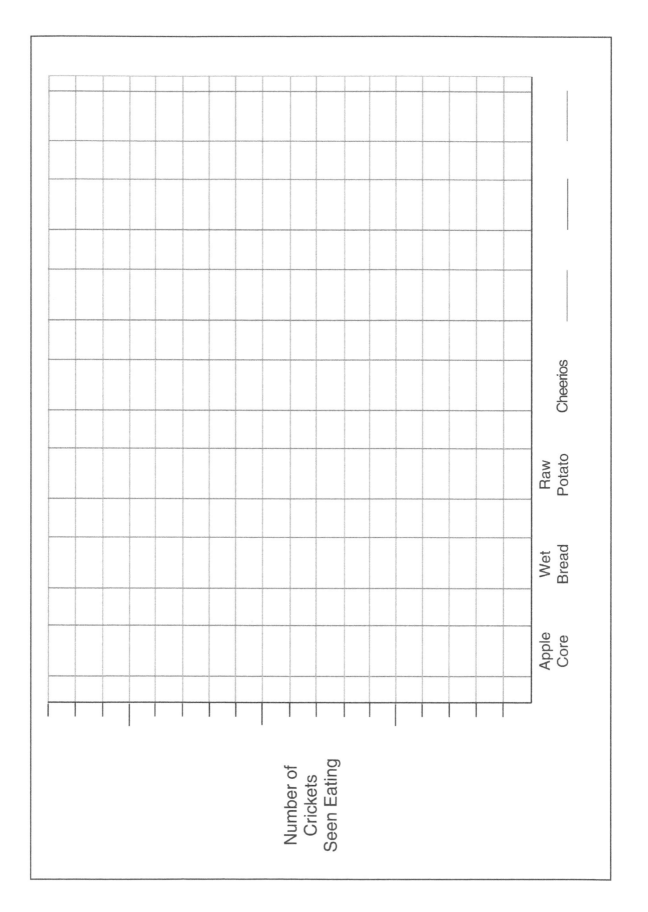

Figure 9.9 Crickets' food preferences in Experiment 8. Fill in the vertical axis according to your sample size. (See Figure 7.4 for an example of a completed histogram.)

to feel more aggressive temporarily, to challenge other male crickets more, and to fight harder. The outcome of any particular encounter depends on which cricket has experienced one or more of those predisposing factors most recently. The next encounter between the same two may have an opposite outcome.

The experiences that increase the level of aggression in male crickets are these: (1) having been isolated for twenty-four hours or more immediately prior to the encounter, (2) having just copulated with a female, (3) having just dominated another male cricket, and (4) having a refuge or territory of his own. Figure 9.10 provides a table for recording how various factors affect aggressive behaviors in Experiments 10 to 13. Figure 9.11 summarizes how various factors affect the general level of aggression, for Experiments 10 to 13. Figure 9.12 is a blank summary chart for the students to fill out.

The male that feels the most aggressive will probably dominate the others. We can manipulate the outcome of a particular encounter between two males by manipulating their circumstances prior to the encounter. Isolate one before the encounter but not the other. Allow one to establish a territory in a refuge (e.g., a matchbox house) and introduce a stranger. Children enjoy trying to predict the outcome of an encounter. I read about the influence of the four factors above before I saw it for myself, and it was fun to actually try each one and see them work.

Experiments 9 to 13 address questions about cricket aggression. Experiment 9 explores general aggressive behavior, while Experiments 10 to 13 explore specific types of aggressive behavior or the effect of one of the four conditions that affect aggression. Some experiments require the students to compare the level of aggressive behavior under these different conditions. Older students may want to assign numerical values to various aggressive behaviors and actually keep a tally in each situation, for comparison. Younger students can make a more subjective judgment about whether or not the crickets are more aggressive in one situation than another.

The following are aggressive behaviors in male crickets: (1) chirping, also called stridulation—the most common aggressive behavior; (2) kicking; (3) raising either head or tail end; (4) head-butting; and (5) wrestling—the most intense of the aggressive behaviors, indicating that the crickets are extremely riled.

Aggressive encounters usually end when one cricket (the loser) retreats or turns around.

Experiment 9

Question:

In a group, is one male consistently dominant over the other males, or does the dominance shift?

Hypothesis:

We think one male cricket will be dominant over the others.

Methods:

You need three to six male crickets. Mark each cricket's thorax with a different color of paint, different position of the paint or different number of dots, and so forth, so that you can tell them apart. (Don't put paint on the abdomen, as they breathe through pores in the abdominal wall.) Isolate the crickets in separate cups for a day or two. Keep a wet shred of bread and a crumpled piece of paper towel in each cup and cover with a cloth secured by a rubber band. Then put all the crickets in a terrarium together, with no refuges for any.

Watch on and off for a few hours (say ten minutes per hour) and record who initiates aggressive encounters and who chirps. (Generally only the dominant cricket chirps.) Do this for two or three days.

Result:

Your result is a statement of your observations of dominant behaviors in each cricket.

Conclusion:

In most cases only one male is dominant in a group, unless more crickets are added or conditions are altered. This means that one male clearly will chirp more and initiate more encounters than the others. If a different cricket chirps the next day, then he has probably defeated the chirper of the day before. If more than one are chirping, then the dominance has probably not yet been established.

In Experiments 10, 11, and 12 the male crickets put in the terrarium first are called residents. The male crickets added later are called intruders.

Experiment 10

Question:

Is a male that has been isolated more aggressive than a male that has not been isolated?

Hypothesis:

A male that has been isolated will be more aggressive.

Methods:

You need two terraria. Small terraria (5 by 3 inches [12.7 x 7.6 cm]) or so are best because the crickets are less able to avoid one another. To get ready, put two males into each terrarium. Give them a day or so to become familiar with the terraria. Meanwhile, keep a fifth male (cricket A), that is marked with paint or otherwise identifiable, isolated for a day or two. Keep a sixth male (cricket B) in a different container with other crickets.

To begin the experiment, add cricket A to the pair of males in one terrarium and add cricket B to the pair of males in the other terrarium. Is the previously isolated intruder (cricket A) more aggressive than the other one (cricket B)? Do the resident males react differently to the isolated intruder (cricket A) than they do to the nonisolated intruder (cricket B)?

Result:

Your result is a statement of your observations of the behavior of cricket A and cricket B and any relevant observations of the other crickets. For recording results, see Figures 9.10 and 9.12.

Conclusion:

In my experience, a cricket that has been isolated is more aggressive. No two crickets will behave exactly alike, but here is an example from a recent experiment I did: Cricket B, the nonisolated intruder, behaved submissively. He ran immediately to a corner of the terrarium where there was a little dip in the sand. Both residents ran immediately to pin him in, leaning over him, trapping him in the corner for fifteen minutes or so. When he finally moved one of the residents chirped aggressively at him. Cricket B did not chirp at all.

In the other terrarium, the isolated intruder (cricket A) was ignored for about fifteen minutes. Then he chirped aggressively at one of the residents that did not chirp back.

To be confident that the results are generally applicable, do several replicates of each situation. You may find somewhat different results.

Experiment 11

Question:

Does having a house make a male more aggressive toward an intruder?

Hypothesis:

A male with a house will be more aggressive toward an intruder than a male without a house.

Methods:

Set up two terraria with one pair of males in each, as described for Experiment 10. Put two houses in one terrarium. A house is anything about the size of a matchbox or a small section of a toothpaste box that a cricket can crawl into. Each house should be open at both ends. Give the males a day or two to get used to their houses. Don't start the experiment until at least one of the males is actually staying in a house. Then add a third marked male cricket to each terrarium. How do the residents react? How does the intruder act?

Result:

Your result is a statement of your crickets' behavior in each situation. See page 141 (column 1) for directions on using Figures 9.10 and 9.12 to record results.

Conclusion:

In my experience, the prediction is confirmed. Male crickets with houses tend to be more aggressive toward intruding crickets than are those without houses. In a recent test I did with several crickets in each group, the housed crickets began to chirp immediately when we put the intruder into the terrarium. Not all chirped, but some did. No crickets in the terrarium without houses chirped for the half hour we watched them. To be confident in generalizing the results to other crickets, you would need to repeat the experiment several times.

Experiment 12

Question:

Does having a female present increase aggression in male crickets?

Hypothesis:

Having a female cricket present does increase aggression.

Methods:

Set up two terraria with one pair of male crickets in each, as described for Experiment 10. Add a female to one terrarium and give the males a day or two to get used to her and to court her or copulate with her or whatever they're going to do. Then add a third marked male to each terrarium. How do the residents react? How does the intruder act?

Result:

Your result is a statement of your observations of the crickets' behavior. See page 141 (column 1) for directions for using Figures 9.10 and 9.12 to record results of Experiments 10 to 13.

Conclusion:

The prediction was rejected in my experience. I have found little aggression in this situation. In a recent test, no one chirped. A resident male kicked the intruder in the no-female terrarium, but that was all. Having recently copulated may increase the aggression of a male, but simply having a female present does not seem to.

Experiment 13

Question:

How does a previous win or loss affect the level of aggression in a male?

Hypothesis:

A cricket that has just won one encounter is in a fighting mood and will probably win his next encounter too. (By "win" I mean that he will dominate the other cricket, chirping more or showing more of the other aggressive behaviors listed earlier.)

Methods:

Set up two pairs of males in two terraria and leave them until one member of each pair is clearly dominant. Put the dominant male from one terrarium (call this one cricket A) with the submissive cricket from the other terrarium (call this one cricket B). Who dominates whom? (A recent win makes a male more aggressive, and a recent loss makes a male more submissive, so cricket A should dominate cricket B.)

Next, match cricket A with other winners. Keep doing this until someone dominates cricket A. Meanwhile match cricket B with other losers until cricket B dominates someone. Now, rematch cricket A and cricket B.

Result:

Your result is a statement of your observations of dominance in cricket A and cricket B. See page 141 (column 1) for directions on using Figures 9.10 and 9.12 to record results.

Conclusion:

Cricket B should dominate cricket A in the end, since he is now the cricket with a recent win. A cricket with a recent win is likely to defeat a cricket with a recent loss.

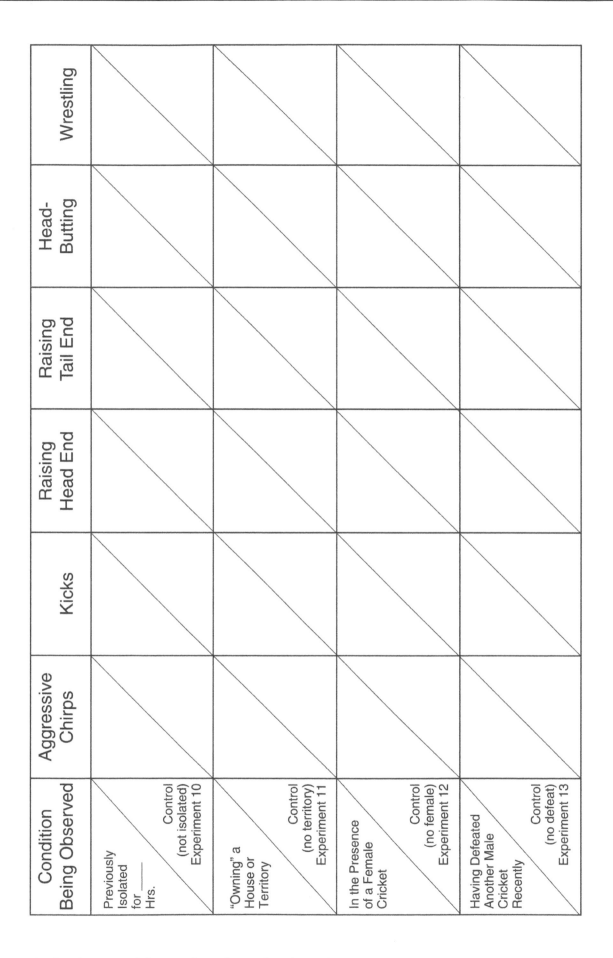

Figure 9.10   Number of aggressive behaviors observed in male crickets in Experiments 10 to 13.

	More Aggressive, Less Aggressive or Neither	**What Aggression Did You See?** (chirps, kicks, raises head or tail end, head-butting, wrestling)
Previous Isolation Experiment 10	More aggressive	Chirps, kicks, raises head end
Owning a House or Territory Experiment 11	More aggressive	Chirps
Presence of a Female Cricket Experiment 12	Neither	None
Having Defeated Another Male Cricket Recently Experiment 13	More aggressive	Chirps and kicks

Figure 9.11 How does each factor affect the aggressive behavior of a male cricket in Experiments 10 to 13? Your results might be different.

	More Aggressive Less Aggressive or Neither	**What Aggression Did You See?** (chirps, kicks, raises head or tail end, head-butting, wrestling)
Previous Isolation Experiment 10		
Owning a House or Territory Experiment 11		
Presence of a Female Cricket Experiment 12		
Having Defeated Another Male Cricket Recently Experiment 13		

Figure 9.12 How does each factor affect the aggressive behavior of a male cricket in Experiments 10 to 13? Your results might be different.

Chapter 10
Many-Legged Millipedes

Introduction

Millipedes are homely, easily overlooked residents of the under-log world. Their most interesting aspects are the patterns of motion in their many legs and their modes of defense, both of which are unusual. They're very low-maintenance creatures for keeping in captivity. One minute of care every two weeks will keep them going, and they're easy and safe to handle. The experiments herein address habitat preference, food detection, response to predators, and more.

After you've done the experiments in this chapter and in Chapters 7, 8 and 9, refer to the table in Figure 1 in the Postscript, which provides a comparison of the results for the experiments in these chapters. Figure 2 in the Postscript is a blank table that can be filled out with the results of your experiments, for comparison.

Materials

1. Millipedes, which can be collected under logs or under the bark of logs or ordered from a biological supply company (see Appendix).

2. A container of some sort for maintaining the millipedes. I use the largest size of plastic margarine tubs, with wax paper lids secured with rubber bands.

3. Soil for the maintenance container.

4. Rotting leaves, rotting wood, and an occasional apple core to feed the millipedes.

5. Two terraria or dishpans or other containers for Experiment 1; one such container for Experiments 2 and 4.

6. A blender for Experiment 2.

7. A cake pan or similar sized container (8 to 9 inches [20.3–22.9 cm in diameter]) for Experiments 3 and 6.

8. A variety of millipede foods (plant parts, fungi) for Experiment 4.

9. A cricket and a praying mantis for Experiment 5.

10. Several other under-log animals for Experiment 6.

Background Information

Millipedes are not to be confused with centipedes. Centipedes can be rather unattractive, especially the brown kind with long legs that you see scuttling around in houses. Millipedes really are not unattractive. They have short, tidy legs that stay more or less tucked under their bodies. And they move much more slowly than centipedes do.

Neither millipedes nor centipedes are insects, although they're both arthropods, which is the same phylum that includes insects, spiders, and crustaceans. Millipedes are no more closely related to centipedes than they are to these other arthropod groups, but they seem so because of the abundance of legs. Millipede means "thousand legs" and centipede means "hundred legs," although neither has that many.

I like millipedes. For several years I've kept a thriving population of them in a big margarine tub—several different kinds of millipedes. They are happy as larks, reproducing like crazy. My captives range in size from as small as a comma to the length of my little finger. There is no larval stage; the hatchlings look like miniature adults except that they are all white and have only three pairs of legs. The adults have many more. As the young ones grow to adulthood, they shed their exoskeletons about seven times, get darker, and acquire more legs.

The defensive behavior of millipedes is interesting. They have two unusual defenses: stinking and coiling. Most millipedes have rows of stink glands along both sides of the body. The glands secrete a foul-smelling and foul-tasting liquid when the millipede is upset. Predators generally avoid them. I once saw a film of an African monkey eating a big millipede during a shortage of other foods and making an awful grimace because of the taste. Most of the millipedes I find are brown to blend in with their background, but some are brightly colored—the colors warn predators. A predator that has tasted another similar millipede will remember the color and avoid it. In my area, a common millipede in summer is black with bright yellow around the edge, called *Sigmoria*. It's pictured, along with others, in the Golden Field Guide, *Spiders and Their Kin*.

Some millipedes also coil when bothered. The head is in the center of the coil, protected. But not all millipedes do this. In my experience the ones that are most cylindrical are most likely to coil. Another millipede defense is seeking cover, although they do move slowly.

How to Tell Millipedes from Centipedes

Any time you see a long, slim insect-like creature with dozens of legs along the entire length of its body, you know it's either a centipede or a millipede. The first things to look for in distinguishing between the two are the shape of its body and the length of its legs (see Figure 10.1). Millipedes are cylindrical, like a drinking straw. Centipedes are flattened cylinders, like you'd have if you laid a drinking straw on a table and flattened it with your hand. Some of the most common millipedes where I live have flattened shelf-like projections sticking out from both sides along the entire length of their bodies. This makes them look somewhat like centipedes (which probably helps them avoid predators since centipedes bite). But there are other ways to tell the two apart. Most centipedes have longer legs than millipedes, and centipedes' legs tend to get longer toward the back ends of their bodies.

Millipedes have short legs, all the same length. Each body segment of a millipede has two pairs of legs so when you look at a millipede from the side, you can see that the legs along that side look paired. Centipedes have only one pair of legs per body segment, so all legs on one side are equally spaced.

Centipedes move with an S-shaped motion. Legs on opposite sides of the body alternate movement. That is, if the first and third legs on the left are moving, then the first and third legs on the right are still. In contrast, most millipedes glide along without wiggling at all. They look as though they are flowing. This is because legs on opposite sides of the millipede's body move in synchrony instead of alternating. One of the activities in this chapter has to do with leg motion.

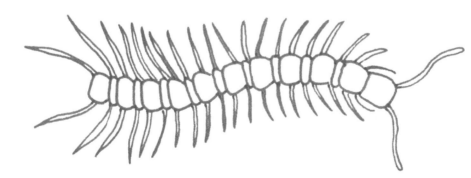

Centipede. Some have much longer legs, such as the common house centipede. The centipede pictured is commonly found under bark on rotting logs. About 2 inches (5 cm) in length.

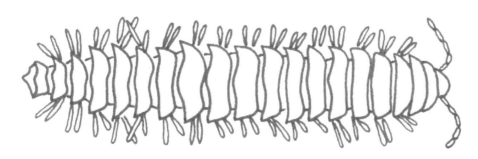

A flat-backed millipede, commonly found in leaf litter. About 1 inch (2.5 cm in length).

Fig. 10.1

How to Get and Keep Millipedes

Centipedes (not millipedes) are difficult to catch. I often see a beautiful orange type of centipede when I pull bark off dead logs. But centipedes run fast and they're gone in a flash. I'm afraid to grab them because they bite. Being predators, centipedes have strong jaws.

Because millipedes move more slowly, they are much easier to catch. They also don't (can't) bite. Millipedes are strictly vegetarians and lack the piercing jaws of centipedes, so you can pick them up.

The best place to look for millipedes is under rocks and under rotten logs, against damp soil. They often burrow down into the soil if conditions are dry. They like damp places. The ones I find under logs are mostly brown and very slim, no longer than my little finger. Most are half that length. I'm more likely to see the brightly colored ones out in the open on the forest floor, unless it's very dry. Millipedes are almost as easy to find as roly-polies. If you turn over several logs and rocks in a damp place, it's almost guaranteed. Don't give up.

Millipedes are very easy to keep. The plastic margarine tub (the largest size) I keep them in is half full of soil. The lid is a piece of wax paper with holes cut into it, secured with a rubber band. A small piece of rotten log and a few damp rotting leaves lie on the surface of the soil, and I throw an apple core in there every couple of weeks. I drizzle some water over the contents of the tub every week or so. I originally set these tubs up for roly-polies, and the millipedes were in there only because a few were clinging to the piece of log I put in for the roly-polies. But the millipedes do as well as the roly-polies. Both have lots of tiny offspring that wander over the soil and fill the nooks and crannies of the log. You can also feed millipedes fungi, the roots of green plants, and seedlings.

Field Hunt

You need an area shaded by trees, where the ground is covered with leaves. Flip logs, rocks, and bricks, and look under the leaves where it's damp. Children can collect them in milk cartons or jars. (Remind them not to stick hands into dark places they can't see.)

If you're unsure about telling millipedes from centipedes, look at pictures online. After you've seen a few, it's quite easy to tell them apart, even when you see a new species. Be sure to provide some moisture for them as soon as you get back to the classroom. A damp crumpled

paper towel will do, while you fix up a terrarium (or margarine tub) for them.

Fig. 10.2 Alan picking up a millipede.

Millipedes at School

Observations and Activities

I first introduce students to the millipedes while they're sitting around a table, one group at a time. I put one to ten millipedes in the middle of the table (see Figure 10.2). All millipedes will start walking if completely exposed like that. The students enjoy watching them walk around, pushing them away from the edge, watching what the millipedes do if someone's finger becomes an obstacle. The next day I have the millipedes available in the margarine tubs, so the students can poke and prod and explore. I sometimes add to this other residents of the under-log world: snails, worms, small slugs, ants, crickets, and so forth. The students can see in the tubs the natural habitat of these creatures in miniature and see the millipedes' natural neighbors. Children seem to especially like baby creatures. They get excited seeing the baby millipedes and baby roly-polies in the margarine tubs. Each student or group can make his or her own under-log world in a container. Children really enjoy fishing around in them (see Figure 10.3).

A good way to start observations of the animals' behavior is to have the students write down everything they notice about the millipedes. Following are some millipede behaviors that I find interesting.

Noticing Millipede Locomotion

The students may notice the wavelike motion of the legs on their own, although I never did before I read about it. If they look at a millipede from the side as it walks, they can see distinct waves of leg motion moving from one end of the millipede to the other. Do these waves of motion move from front to rear or from rear to front? (Rear to front.) How many waves can the students see at one time? On the millipedes I've watched I can see three to four waves moving at once.

Fig. 10.3 Alan getting millipedes out of his margarine container to put in a petri dish.

If the students look carefully, they can see that what appears to be a wave is really the legs that are raised off the surface. Most of the legs are in contact with the ground at any one time. Only a few are raised. At the beginning of the wave, legs at the back of the body are raised, then as they are put down, those just ahead of them are raised and so on. Thus, the raising and putting down of legs looks like a ripple or wave moving from the rear forward along the legs. Since millipede legs on opposite sides move in synchrony, the two sides look identical at any one time.

Millipede Grooming

What will a millipede do if it gets something goopy on its legs? Set the millipede on a surface smeared with a thin layer of jelly, honey, or vegetable oil. What does it do? Millipedes will clean and groom all their legs with their mouthparts when they get sticky or dirty. Millipedes are like cats and mantises in a way, using their mouths to clean their appendages.

Millipede Defenses

The students may notice on their own that the millipedes smell bad. The little brown ones usually don't smell as bad as the larger and brightly colored ones. students may have to put the little brown ones up close to their noses to smell them or sniff their fingers after handling them. I can smell it at a distance, maybe because I know what it smells like.

Larger ones can be smelled at arm's length. The brightly colored *Sigmoria* activates its stink glands at the slightest provocation, before the little brown ones do. What line of defense does your millipede use first: running, coiling, or stinking? If you have more than one species, the students may notice that some coil and some don't. Put the millipede on a flat surface and watch it. After a minute or so, poke it with something soft, like a paintbrush, or touch it gently with your finger.

Different types of millipedes may react differently. In my experience, the little brown cylindrical ones that are so common in my area will run as a first defense, will stink if annoyed further, and will coil as a last resort. Those with projections on the sides, however, don't coil. Millipedes don't usually use all their defenses at once. Why not? If they discharged the contents of the stink gland every time they saw something moving, they'd spend an awful lot of energy synthesizing new fluid for the glands. It makes sense too that coiling should be a last resort, because it doesn't protect them from large predators who might eat them whole, but only from smaller ones that might go for a piece of a millipede.

Experiments

Remember that the hypotheses I give are just examples. Your hypotheses will be the predictions made by the class or a particular student. Your result for each experiment will be a statement of how your animals reacted to your experimental setup. Your conclusion is a statement of whether your prediction was confirmed or not. For each experiment, adding replicates increases your confidence in the validity of your conclusion, but they may be omitted if tedious for young children.

Experiment 1

Question:

Can millipedes find their food as quickly in the dark as in the light?

Hypothesis:

Millipedes need light to find their food.

Methods:

Catch several millipedes, preferably of the same type. Divide them into two groups of equal number. Ten millipedes in each group is optimum, but fewer is okay. Provide moisture but no food to each group for three or four days. Meanwhile, get empty containers of equal size, about 12 to 20 inches (30.5–50.8 cm) in diameter, like plastic dishpans or baking pans. Put millipede food in a small pile in the center of each tub: plant rootlets, a seedling, a piece of fruit with mold on it, and/or pieces of rotting leaves. After three or four days without food, put the millipedes in one corner of each box. Cover one box so that it's completely dark inside. Leave the other open. Wait three minutes, then count how many millipedes are on the food pile in each container. If none are, wait longer. Did more millipedes find the food in the light than in the dark?

Result:

Your result is a statement of the number of millipedes on the food in each box. You can record your results on Figure 10.4, a histogram. Figure 7.4 provides an example of a completed histogram.

Conclusion:

The prediction is not confirmed. The number of millipedes on the food should be about the same in the two containers. They locate their food by scent and have poor eyesight. Since millipedes are nocturnal, it makes sense that they would rely more on their sense of smell.

How else could you test this? See the next experiment.

Experiment 2

Question:

Can millipedes find food by scent alone?

Hypothesis:

Millipedes need to be able to see food to find it.

Methods:

Put some seedlings, rotten leaves, fungus, and plant roots in a blender with some water and grind it into a liquid. Soak a small piece of absorbent cloth with the liquid. Soak another piece of absorbent cloth with water. Put the two cloths on opposite sides of a container and place your millipede(s) in the middle of the container. Which cloth is chosen, if either?

Result:

Your result is a statement of your observation.

Conclusion:

The prediction is not confirmed. Millipedes are attracted to the blender liquid by scent, but not the water.

Experiment 3

Question:

What types of conditions do millipedes prefer?

Hypothesis:

They like a bright, sunny place.

Methods:

You can use one or several millipedes here. Offer the millipede(s) a series of choices. Just about any container big enough for the millipede to move around in will do, but it must have low sides. A cake pan is a good size (8 to 9 inches [20.3–22.9 cm] in diameter), with a Plexiglas or tight plastic wrap lid. First make one side damp and one side dry by covering the bottom of the cake pan with two semicircles of construction paper, one damp and one dry. The two pieces of paper shouldn't be touching or the dry piece will soak up dampness. Put one or several millipedes in the center of the dish and leave them for a half hour or so. Which side do they prefer?

Now make one side dark and one side light by covering the top on one side with something dark and opaque, like a piece of black construction paper. The bottom should be uniformly damp or uniformly dry. Put the millipede(s) in the center and come back in a half hour or so. Which side do they prefer? Does the dampness of the substrate affect the outcome?

Make a little house in the pan with a small box or a section of a paper towel roll cut in half lengthwise. Leave the house with the millipede(s) for a half hour or so. Do they hide in it or stay out? Does the dampness of the substrate affect the outcome?

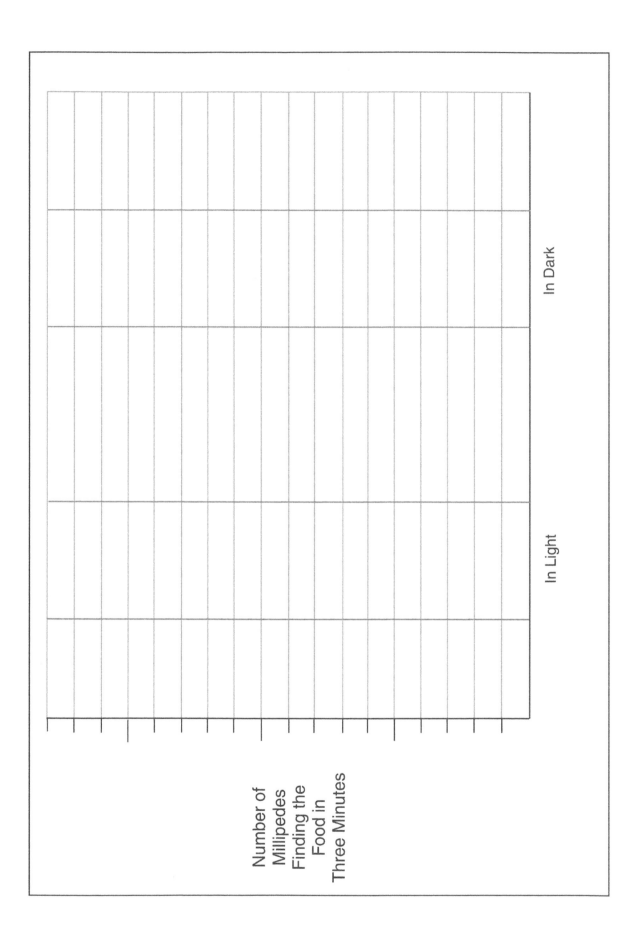

Figure 10.4 Blank graph for recording your results for Experiment 1. (See Figure 7.4 for an example of a completed histogram.)

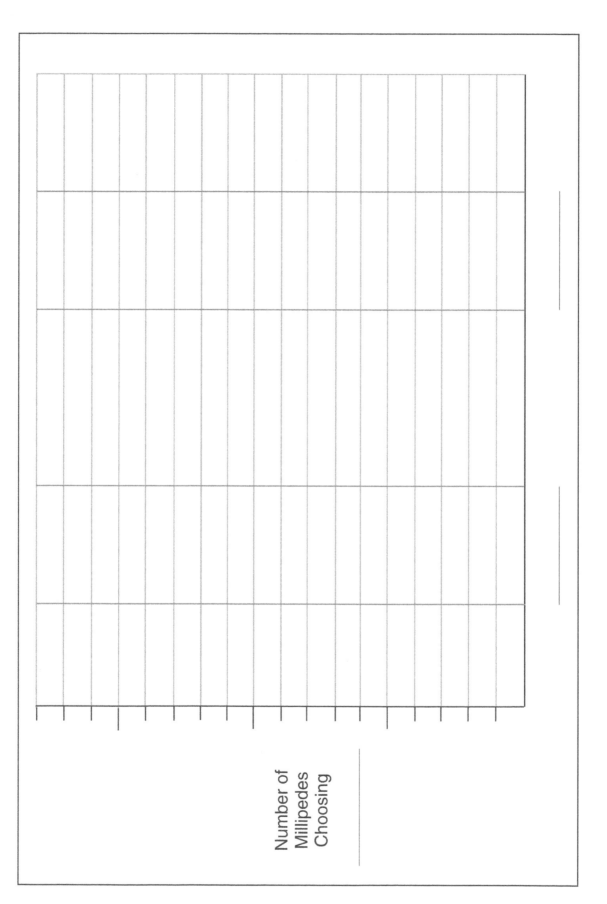

Figure 10.5 Blank histogram for recording your results for Experiment 3. (See Figure 7.4 for an example of a completed histogram.) Fill in the blanks on the horizontal axis with these pairs of labels, for the four sets of conditions offered in Experiment 3: Dampness-Dryness, Darkness-Light, Cover-Open, Warmth-Cold.

Choice of Conditions Offered to Millipedes	Millipedes' Choice	Effect of Choice on Millipedes
Dampness or Dryness		
Darkness or Light		
Cover or Open Space		
Warmth or Cold		

Figure 10.6 Blank table for recording your results for Experiment 3.

Get a long cookie pan with sides. Put a hot water bottle under one end of the pan. Put a tray of ice under the other end. After waiting awhile for the respective ends of the tray to heat and cool, put the millipede(s) in the middle. Come back after a half hour to an hour. Do they prefer the warm end or the cool end or the middle?

Result:

Your result is a statement of your millipede's choices. You can record your results on Figure 10.5, a histogram. Figure 7.4 provides an example of a completed histogram. You can also record your results in the table provided in Figure 10.6. Figure 8.8 is a similar graph, for earthworms.

Conclusion:

The outcome probably depends on the type of millipede. Those found under logs will tend to prefer darkness, dampness, cover, and coolness. We see millipedes under logs because they find the conditions they prefer there.

Experiment 4

Question:

What foods do your millipedes prefer?

Hypothesis:

It depends on what you offer.

Methods:

Use several millipedes for this one if you can. Put damp paper on the floor of a container about the size of a dishpan. Otherwise the millipedes may spend all their time trying to escape instead of looking for food.

Offer your millipedes a choice between small seedlings (plants that have just sprouted from seeds), fungus (like the mold that grows on a rotten fruit or a mushroom), small rootlets of plants, and rotting leaves. Put the four choices in different corners of a container, and after a half hour or longer, record which one your millipedes prefer.

Result:

Your result is a statement of your millipedes' choice. You can record your results on the table in Figure 10.9.

Conclusion:

The outcome depends on what species of millipede you have.

Experiment 5

Question:

Does stinking deter predators?

Hypothesis:

We think a predator will avoid a millipede that stinks.

Methods:

Offer a praying mantis a millipede and a small cricket of comparable size. Make sure both are moving. Which does it take?

Result:

Your result is a statement of your mantis's choice.

Conclusion:

The prediction is confirmed. The predator will probably avoid the millipede.

Experiment 6

Question:

Do other under-log inhabitants avoid millipedes?

Hypothesis:

We think other under-log inhabitants will avoid millipedes, especially millipedes we can smell.

Methods:

This experiment involves comparing (1) the distance between a millipede and another animal to (2) the distance between two things that we know to be indifferent to each other, like two peas. To do this you'll need a small container with steep sides, such as a cake pan. Draw a grid on the floor of it with at least nine squares in the grid. Put two peas in the pan, shake it and put it down without looking. Then count the number of lines you'd have to cross to get from one pea to the other without curving.

If the peas are in the same square, that counts as zero. If they are in adjacent squares, that counts as a one. If they are in squares that touch diagonally, that counts as a two, and so on. Do this ten to twenty times with a pair of peas and calculate an

FOODS ACCEPTED BY MILLIPEDES			
	Accepted	Not Accepted	Not Sure
Seedlings			
Fungus			
Rootlets			
Rotting Leaves			
Raw Potato			
Chunk of Apple			
_____			
_____			
_____			

Figure 10.7 Blank table for recording results for Experiment 4.

average. This is the number of lines you would expect to be between two animals if they were totally indifferent to one another.

Now do the same thing with pairs of animals, except there's no need to shake them. Put one millipede and one other animal at a time in the container and cover it with a clear lid like plastic wrap taped down tightly. (The animals will be able to breathe adequately for the short period of the trial.) Handle or prod the millipede first to try to get it to discharge its stink glands.

Here are some suggestions for animals to pair with the millipede: a worm, a snail, a slug, a roly-poly, a beetle, a spider, and a centipede. Check the pair after an hour. Record how many lines you'd have to cross to get from one animal to another without curving, as described for the peas. Do this ten times for each pair of animals. Then calculate an average number of lines for each pair, as you did for the pair of peas. Are the animals farther apart in general than the peas were? This is what you'd expect if the nonmillipedes are avoiding the millipedes. Here are some ways to vary the experiment:

1. Use a millipede and one pea instead of two peas to calculate your average for two things that are indifferent to each other.

2. Try pairs of animals where neither is a millipede and compare.

3. Try pairs of millipedes and compare.

4. Make your grid on a piece of brown paper towel and dampen it. Compare results on the damp substrate to results in a dry dish. Danger of dehydration may cause animals to seek out other bodies.

Result:

Your result is a statement of your average line counts for pea pairs, for millipede-plus-other pairs and for pairs of nonmillipedes and pairs of millipedes, if you go that far. You can record your results on the table in Figure 10.8.

Conclusion:

The prediction is confirmed generally. Other animals may avoid the millipede, especially a very stinky one. A millipede may seek out another millipede to mate, but they are otherwise generally indifferent to each other. During mating, the male lies on top of the female, their bodies aligned. He also nibbles at her face.

Other small invertebrates are often indifferent to one another unless they are dehydrating. Huddling together or "bunching" reduces the amount of surface area in contact with drying air. I've never seen millipedes do this but they may in a dry dish. It would make sense for under-log residents to do it, because they are adapted to a damp environment and probably vulnerable to dehydration.

Do other under-log inhabitants avoid millipedes?

Number of lines between peas (shake and count ten to twenty times). Record your counts here:

What was the average number of lines between peas?

Number of lines between millipede and worm (count ten to twenty times), allowing time for animals to move before counting again. Record your counts here:

What is the average number of lines between millipede and companion?

Repeat with millipede and roly-poly.

    Record counts here:

    Average number of lines:

Repeat with millipede and

    Record counts here:

    Average number of lines:

Were the pea pairs closer than the animal pairs?

Figure 10.8   Sheet for recording results for Experiment 6.

Chapter 11
The Slime That Creeps

Introduction

A slime mold is a small, goopy organism that stumped taxonomists for a long time because it is in some ways like a fungus, but also has some animal-like characteristics. It reproduces with spores, like a fungus, but it can creep slowly like an amoeba and pursue its food!

Slime molds are easy to maintain and fascinate children because they are so peculiar. This chapter describes how to keep a slime mold in captivity and how to conduct experiments to answer questions about its feeding, its reproduction, and its ability to regenerate.

Materials

1. Slime mold plasmodium. This can be cultured from dead wood, which may take a couple of months, or ordered from a biological supply company (see Appendix).

2. Petri dishes and lids, which can be bought at a science hobby shop or ordered from a biological supply company. One dish per student works well, or the students can work in groups. If you want to do all of the activities in this chapter, you'll need at least two dishes for each student or each group. Four-inch (10.2 cm) diameter dishes are a good size.

3. Filter paper. You need at least one sheet per dish, the same diameter as the dish. You should have a few spares.

4. Tweezers or forceps.

5. Oatmeal or other food, if desired.

Fig. 11.1 The plasmodium of a many-headed slime streaming or "creeping" up the side of the petri dish. Many have "escaped" onto my kitchen counter in this fashion. Diaper pin added for scale.

Background Information

A slime mold is not particularly charming at first glance. It looks rather like something you'd find in a jar of leftovers that's been in the back of the refrigerator too long. But it's different from other molds, which digest and absorb the surface they're growing on. The difference is that slime molds are able to move about freely and "capture" their food. This odd feature—the ability to creep—is what first attracted my attention to them.

A slime can be kept in captivity indefinitely as long as you feed it regularly, keep it damp, and are careful not to introduce bacteria or other molds. When you feed your slime an oatmeal flake, it will creep slowly toward the flake to engulf it, which is wonderfully fun to watch! Slimes are like clocks in that you can't see them move, but you can easily see that they have moved over the span of a few minutes. Slime molds have a two-part life cycle. The two phases don't occur at the same time but rather alternate. The phase that creeps after its food is called a plasmodium (see Figure 11.1). The plasmodial phase of all slime molds prefers darkness and moisture. In nature they stay under or inside rotten logs and are not easy to find. I have found active plasmodia in nature only a few times, usually in damp logs near streams and once on the ground next to a hiking trail.

The other phase of the slime life cycle, the reproductive phase, is much more conspicuous in nature than the plasmodium. During the reproductive phase all slimes make some kind of little capsule containing spores. (A spore is like a tiny seed except that it usually isn't produced by sexual union as a seed is but rather by a single organism.) These reproductive capsules are easier to find than the plasmodium because they're often in the open, such as on top of a log. On top is a better place for the slime to disperse its spores. The reproductive capsules are called sporangia, and all fungi make sporangia (see Figure 11.2).

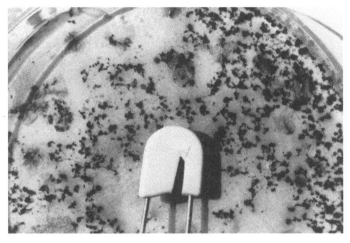

Fig. 11.2 Sporangia, or the reproductive phase, of the many-headed slime in a 3.5 inch (9 cm) petri dish. Diaper pin added for scale.

The slime species most often kept in captivity is the many-headed slime, which has a bright yellow plasmodium. When pursuing a food item, the slime stretches itself out in long streams of yellow protoplasm (see Figure 11.1), but while eating an oatmeal flake, it looks like a yellow blob of goo. The slime mold doesn't actually creep; it really streams. It sends out a membrane in the shape of a channel, and then its protoplasm streams into the channel. We have protoplasm too! People can't stream from place to place because our protoplasm is all contained in cells bound by cell membranes. But the slime mold's protoplasm isn't packaged, so it flows along from one place to another.

How to Get and Keep Slime Molds

Growing Plasmodia from Spores

Any piece of rotting log will probably have spores on it that will often germinate if the wood is kept damp for a long time. I did it by keeping a wet washcloth over a piece of rotting log in a terrarium for two to three months. I moistened the washcloth every day. After two and a half months, a many-headed slime plasmodium appeared, followed by a cluster of carnival candy slime sporangia.

For some reason, plasmodia on logs are much more hardy and resistant to contamination by other molds or bacteria than are those in petri dishes. Perhaps the slimes have an advantage on the log because the log is their natural habitat and they are finely adapted to it. I have a many-headed slime plasmodium now that lives on a damp piece of rotting wood in a terrarium. It's one that I ordered and then released onto the log. I never feed it or tend to it in any way, other than keeping a damp cloth over the piece of log.

A log is ideal for keeping one as a pet or as a "stock" slime for the long term. You can start one in a dish by taking a piece of the one on the log, although it may disappear into the log periodically and be inaccessible.

Ordering a Plasmodium

If you order a many-headed slime mold (*Polycephalum physarum*) from Carolina Biological (www.carolina.com), or from Science Kit (sciencekit.com), you can get an active plasmodium in a petri dish (a "plate culture"). See their websites or give them a call for more details. Contact info is in the Appendix of this book.

Propagating a Plasmodium

If you order an active plasmodium and want to propagate it for a class, the best thing to use is petri dishes. If you want to avoid the expense (which isn't much), kitchen bowls or saucers will do. You'll need to keep them tightly covered with plastic wrap to keep moisture in. Whether you use petri dishes or kitchen bowls you have to have absorbent paper in the bottom of the dish. Filter paper works well, as it holds water for about twenty-four hours. You can order a box of one hundred papers the same width as your petri dishes.

To get the slime from the plate or tube into a new dish, you can just scrape some up or cut into pieces the rubbery gel it comes on. Put a small glop of slime or a piece of the gel with slime on it into the new dish. Take the gel out when the slime creeps off of it. It's best to use more or less sterile tweezers and scissors.

If you scrape up the plasmodium to transfer it, it will take a day or two to recover and begin streaming again.

Maintaining a Plasmodium

To maintain a plasmodium in one dish, feed it no more than one oatmeal flake a day (see Figure 11.3). If you feed it more the food will rot. Add drops of water as needed to keep the filter paper barely moist. I use tap water that has been left standing uncovered twenty-four hours for the chlorine to vaporize.

Fig. 11.3 Pinching up one more oatmeal flake to feed the many-headed slime. Someone has put in too many! The tweezers help minimize contaminating the slime with bacteria and the spores of other molds.

The instructions that come with the slime say that everything that touches the slime must be sterile, including the water. (Boiling the water works.) I've had mixed experiences with this. Sometimes I can handle the filter paper or oatmeal and somehow not contaminate the culture. But if you let children add the oatmeal and touch the slimes with unwashed hands, you'll probably get a vast array of molds and other growths to keep the slime mold company. Once a particular plasmodium is contaminated, it's almost impossible to separate it from the other growths, and it usually dies.

Field Hunt

It's a lot of fun to find slime molds in nature, and they're quite common. It takes practice to be able to spot them because they are usually very small. Even the brightly colored sporangia are relatively inconspicuous because of their size. The first time I found a slime in nature I didn't know what it was. It looked like a reddish dusting of powder on a rotten log. Scattered throughout the fuzz were clusters of tiny black cups, each cup about the size of the head of a pin. I wondered if it might be a fungus because it was obviously something living, but it wasn't a plant or animal. I flipped through Audubon's mushroom field guide when I got home and found it easily, because it looked exactly like the color photograph. It was multigoblet slime. What a name! That began my interest in slimes. (Slimes are no longer classified as slimes, so are probably not in newer editions of fungus field guides.)

The sporangia of each slime species are distinctive. It isn't hard to tell them apart with pictures. You can search the Internet for images for any of these names and find good pictures. Chocolate tube slime sporangia look like little brown cigars on stalks; carnival candy slime sporangia look like tiny pink puffs of cotton candy on stalks; scrambled egg slime sporangia look like scrambled eggs; yellow-fuzz cone slime sporangia look like tiny cones, some of which are full of yellow fuzz (see Figure 11.4). I've found all of the above in nature, most of them when I was just walking and not really looking.

Fig. 11.4 Sporangia, or the reproductive phase, of the yellow-fuzz cone slime in nature.

Now that I have shown my children several in nature and the pictures too, they've become pretty good at finding sporangia by themselves. Their favorite is toothpaste slime, also called wolf's milk slime. It looks like a cluster of round pink pillows, each about the size of a green pea (see Figure 11.5). My children like the toothpaste slime sporangia partly because we've found them several times and partly because if you squeeze one gently, it bulges and then ruptures with a satisfying pop.

Out oozes a deep pink paste, somewhat like the consistency of toothpaste. Of course, the kids find popping them irresistible, squealing, "Ooh, gross!" and "Aah." They love it. (Okay, I do too.) It's fortunate that they occur in clusters—we never pop more than a few, leaving the majority to mature.

Left alone, the paste in toothpaste slime will dry out and turn into a packet of thickly packed dust. The dust is actually spores, encased in a little round sack with a pinhole at the top. Touch it and a cloud of pinkish "smoke" comes out.

Fig. 11.5 Sporangia, or the reproductive phase, of toothpaste slime (sometimes called wolf's milk slime) in nature.

Slimes at School

Getting Ready

After you've gotten as many slimes as you want growing in individual dishes, you're ready to take the dishes to school. Don't wait too long between getting the dishes ready and taking them to school, because if you have introduced molds or bacteria to the dish, they may overwhelm the slime in a week or so.

Observations and Activities

Feeding the slime molds is a good place to start, whether you want to do only that or go on to an experiment. After a few introductory remarks to the students describing what a slime is and what is interesting about it, pass out a dish of slime and a flake of oatmeal to each child. You should probably put each oatmeal flake on a small piece of paper so that the students can dump the oatmeal flakes into the dishes by picking up the papers without touching the flakes. They can use clean hands

instead if you don't plan to keep the slimes more than a week or so. Tell them to put the oatmeal close to their slimes. Some of the oatmeal flakes will be engulfed by the slimes right away, but it may take a couple of hours before all are. The bright yellow color of the many-headed slime makes it easy to tell when it has streamed onto the oatmeal. Help the students describe what they've seen.

You can extend these feeding observations and help the students begin to feel involved in decision making by asking them what other kinds of things they think the slimes might eat. Encourage them to bring in their suggestions. They may bring food from home or things from nature that aren't food for us but could be for the slime mold. You may want to encourage them to think about what the slime mold feeds on in its natural habitat of rotting logs. (Many feed on microorganisms in the log.) They digest a surprising variety of foods.

Food remnants in the dish seem to encourage the growth of bacteria and other molds that can be lethal to the slime mold, so you should probably set aside a couple of dishes for food testing and not plan to use them for anything else. Or the students can feed their own slimes the different foods if they do it after the other activities have been completed or if they all have extra slimes.

During their observations, the students will probably come up with questions. Here are some of the questions I've been asked by first graders watching the slimes eat: You mean it's blind? Could it use glasses? Why doesn't it have any head? How does it know where to go? Does it have teeth? Does it have a tongue? Does it have a mouth? How does it eat? It doesn't have any senses? If they don't have a head, how do they know where their food is? If they don't have a head, how do they know anything?

Children seem to be fascinated by how something that possesses the ability to move independently can be so devoid of other similarities to ourselves. Slimes get by just fine with only one sense—some sort of taste or chemical detection. Get the students to think about whether slimes really need any other senses. (This is a good opportunity to start them thinking about how and why other animals are different from us. Does a tree need to be able to see or hear? How about a clam or an earthworm? No, because they are all either sedentary or burrowing.)

Another avenue of exploration is the ability of slimes to regenerate, which you've already seen yourself if you divided the one you ordered to make more for the students. If you have enough petri dishes you can let

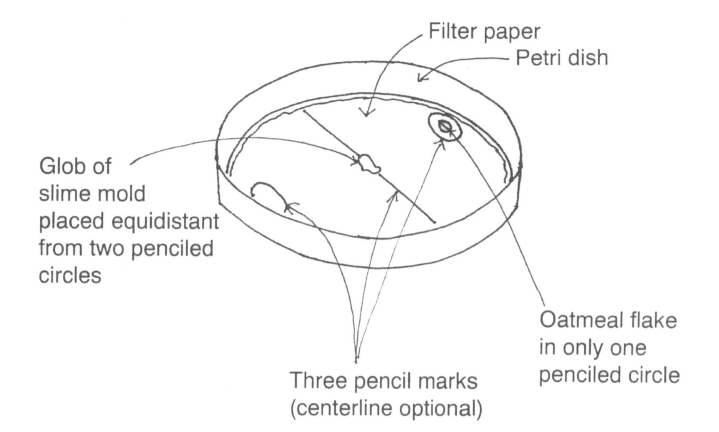

Filter paper

Petri dish

Glob of slime mold placed equidistant from two penciled circles

Oatmeal flake in only one penciled circle

Three pencil marks (centerline optional)

Fig. 11.6 Petri dish setup for Experiment 1.

each student use a toothpick to scoop up the slime and transfer it to another dish. Feed the new one. How long does it take the new one to begin streaming toward its food? How small a portion of slime can they transfer to a new dish and still have a new slime develop?

Seeing the slimes survive and grow after being cut apart seems to intrigue children. They ask a lot of questions about this too: Why doesn't it kill them? How do you know it doesn't hurt? Why doesn't it hurt?

This is a good time to explain to them that the slime has no different body parts like we do. It has no organs inside, no heart, no stomach, no arms, no legs, no nerves. Every piece is, in a way, a whole body.

Experiments

Remember that the hypotheses I give are just examples. Your hypotheses will be the predictions made by the class or a particular student. Your result for each experiment will be a statement of how your fungi reacted to your experimental setup. Your conclusion is a statement of whether your prediction was confirmed or not. For each experiment, adding replicates increases your confidence

in the validity of your conclusion, but they may be omitted if tedious for young children.

Experiment 1

It's a good idea if you can derive your experiment from the students' questions. Sometimes the opportunity doesn't present itself, but for me it did with the slime molds. One child asked, "If they don't have a head, how do they know where the food is?" "How do they know" is not the type of question that lends itself to being answered by an experiment, but it can be modified slightly so that it is. The question, "Do they know where their food is?" is one that can easily be answered by an experiment. You can answer it the same way you would for any animal. Offer them food at a distance and see if they move toward it faster than they move toward a designated spot with no food. That's what we did.

Question:

Do slime molds know where their food is or do they just stumble upon it randomly? Specifically, will they move faster toward their food than toward a designated spot with no food?

Hypothesis:

We think the slimes just randomly encounter their food.

Methods:

You'll need at least about ten individual slimes to participate. The students may volunteer theirs. The more you have, the more confidence you can have in your results. You'll need a fresh petri dish for each participant. (You can wash and reuse them. Let them air-dry to help kill bacteria.) Before you put the filter paper in the dish, while the paper is still dry, draw two circles a little larger than an oatmeal flake on opposite edges of the filter paper. Then put the paper in the dish and moisten it. Put an oatmeal flake in only one of the circles, and leave the other one empty. Place a wad of slime in the center of the paper (see Figure 11.6). Check it every half hour or so and record which circle the slime enters first. If it enters the circle with the oatmeal-flake paper first in most of the dishes, then it probably "knows" where the oatmeal-flake paper is. If it enters the empty circle first in about half of the dishes, then you can conclude that its movements are random and it is not moving purposefully in one direction.

Result:

Your result is a statement of how many slimes reached the food first, how many reached the empty circle first, and any other relevant observations. You can record your results as a histogram in Figure 11.7. Figure 7.4 provides an example of a completed histogram.

Conclusion:

In my experience, the slimes will start out moving randomly, sending out streams toward both circles. When one stream of protoplasm comes to within about $1/2$ inch (1.3 cm) of the oatmeal-flake paper, then the rest of the plasmodium begins to stream toward it too. Apparently they somehow detect in the filter paper chemicals (carbohydrates, proteins) that have diffused into the paper around the oatmeal. My results have been that most of the slimes enter the circle with the oatmeal flake first, but not all of them do. But don't tell the students this until after the experiment is over or there'll

be no suspense! Besides, you may get a different outcome. In my experience, this prediction has been rejected. Slime molds can move purposefully toward their food.

Experiment 2

How would you determine more definitively the distance at which the slime can detect the oatmeal? You could vary the distance between slime and oatmeal in different dishes. You would record how soon the oatmeal ceases random movement and begins to stream purposefully toward the oatmeal in each dish.

Question:

At what distance can the slime mold detect the oatmeal in the filter paper? In other words, at what distance will most of the slimes begin to move definitively toward the food right away?

Hypothesis:

The slime mold will move all of its protoplasm toward the oatmeal right away when it is introduced at a distance of 1 inch (2.5 cm) (or whatever distance your students predict).

It will move in several different directions initially when deposited at a distance of 3 inches (7.6 cm) (or whatever the students predict) from the oatmeal.

Methods:

Vary the initial distance between slime and oatmeal from $1/4$ inch to 3 inches (.6–7.6 cm) in different dishes. At what initial distance do the slimes begin to move all of their protoplasm toward the oatmeal right away?

Result:

Your result is a statement of the slimes' behavior at different initial distances. You can record your results on the table in Figure 11.8.

Conclusion:

In my experience the distance at which the slimes detect the oatmeal and begin to move all of their protoplasm toward it is between $1/2$ and 1 inch (1.3–2.5 cm).

I'm not sure what causes the slime mold to shift from one stage of its life cycle to another. Scientific explanations include that exposure to light, lack of

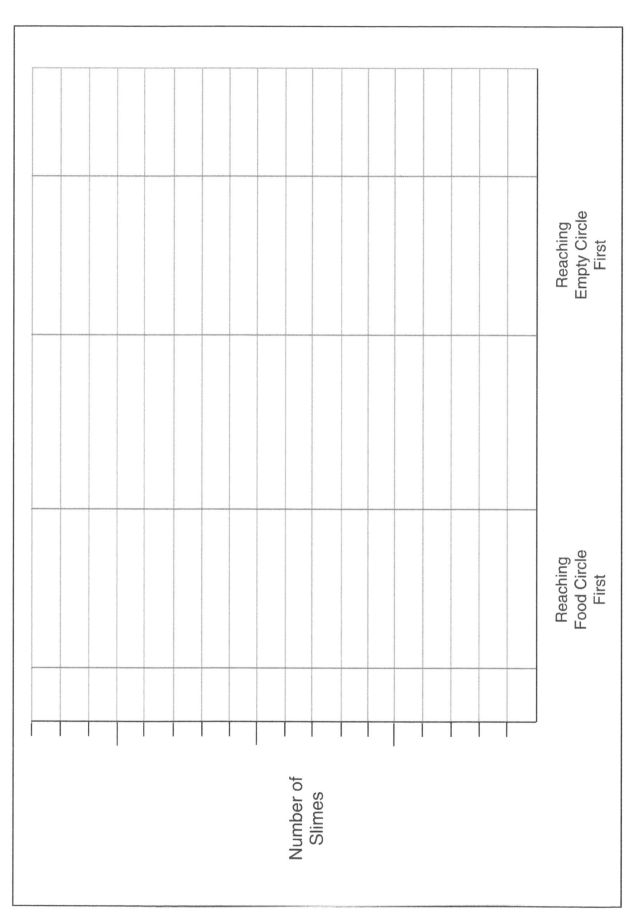

Figure 11.7 Blank table for recording results for Experiment 1. (See Figure 7.4 for an example of a completed histogram.)

Initial Distance	Did All of the Protoplasm Move Toward the Food and Not in the Other Direction?	
	Yes	No
¼ in.		
½ in.		
¾ in.		
1 in.		
1¼ in.		
1½ in.		
1¾ in.		
2 in.		
2½ in.		
3 in.		

Figure 11.8 Blank table for recording results for Experiment 2.

moisture, or absence of food can cause sporangia to form from the plasmodium. At least one of these three conditions has been present every time I've seen sporangia form. It often happens when I stop taking care of them. If I throw a plasmodium in the trash while I'm cleaning some petri dishes, the next morning I'll probably see sporangia in the trash. Or, if a plasmodium escapes from its container, which often happens, the escaped portion often forms sporangia. This happens as a way of self-preservation and reproduction when conditions are not right for the plasmodium. Contact with plastic wrap seems to cause sporangia to form immediately, maybe because the plastic wrap is so completely nonabsorbent and devoid of moisture. I've put filter paper with slimes aboard on plastic wrap and found the next day that the edges that had crawled off the filter paper had formed sporangia, while the part remaining on the filter paper was still plasmodium.

Other times I've had slime molds dry up completely and not form sporangia. This uncertainty about what exactly causes it can lead to some interesting experiments. If the students know that plasmodia are normally found inside rotting logs, get them to talk about what it's like in there: dark and damp (and full of organic matter, although the children probably won't come up with this one). Sporangia form on top of the log. What's it like on top? Light and probably drier.

Help the students speculate about what conditions the plasmodium may need. Moisture and darkness and food. A table is provided in Figure 11.9 to record which factors cause sporangia formation. How can you test which of these factors will cause sporangia to form? Remove each factor one at a time. Let the students volunteer their slimes, although you may find that none want to volunteer, for fear of losing their plasmodium. If not, maybe you can use some spares for this experiment or give the students replacements.

Experiment 3

Question:

Will drying induce sporangia to form?

Hypothesis:

We think drying will induce sporangia to form.

Methods:

Stop watering some of the slimes, while continuing to feed them moistened oatmeal flakes and keeping them in the dark. Continue to water some of the slimes, as a control.

Result:

Your result is a statement of your observations. You can record your results on the table in Figure 11.9.

Conclusion:

My experience is that some of the slimes that are not watered form sporangia but some don't; some merely dry up. Thus, allowing the filter paper to dry may cause sporangia to form but not always.

Experiment 4

Question:

Will light cause sporangia to form?

Hypothesis:

We think light will cause sporangia to form.

Methods:

Leave some in the light, some in the dark. Otherwise care for both groups with the usual feeding and watering.

Result:

Your result is a statement of your observations. You can record your results on the table in Figure 11.9.

Conclusion:

In my experience, light can but does not usually cause sporangia to form.

Experiment 5

Question:

Will lack of food cause sporangia to form?

Hypothesis:

We think lack of food will cause sporangia to form.

Methods:

Stop feeding one group of slimes, while continuing to feed the other. Otherwise maintain both groups in the dark and with adequate moisture.

Result:

Your result is a statement of your observations. You can record your results on the table in Figure 11.9.

Conclusion:

In my experience, lack of food can but does not always cause sporangia to form.

Experiment 6

Question:

Does placing or allowing the slime to crawl onto a completely nonabsorbent substrate like plastic wrap cause sporangia to form?

Hypothesis:

We think being put on plastic wrap will cause the slime to form sporangia.

Methods:

Put the slime on plastic wrap or allow a portion of it to crawl onto plastic wrap.

Result:

Your result is a statement of your observations. You can record your results on the table in Figure 11.9.

Conclusion:

In my experience, being on plastic wrap usually does cause sporangia to form.

Experiment 7

Remove any combination of food, darkness and moisture and see what happens.

Which Conditions Caused Sporangia to Form? Put a Check in the Appropriate Space.			
	Drying	Light	Lack of Food
Slime 1			
Slime 2			
Slime 3			
Slime 4			
Slime 5			
Slime 6			
Slime 7			
Slime 8			
Slime 9			
Slime 10			

Figure 11.9 Blank table for recording results for Experiments 3 to 7.

Terrestrial Predators

Chapter 12
Ant Lions:
Terrors of the Sand

Introduction

Most of the children I've worked with have never seen or heard of an ant lion, but they are soon fascinated by these creatures. In this chapter I describe where to find ant lions (sandy playgrounds mostly) and how to set them up in the classroom so that they'll build their spectacular prey traps, which are pits in sand. Ant lions are insects, but they're similar to spiders in their feeding methods. I also describe several experiments regarding the ant lions' pit-building behavior. Ant lions require very little maintenance—an ant or other small insect for dinner every other day or so.

Figure 3 in the Postscript provides a range of questions and answers comparing behavior of the predators discussed in Chapters 12 to 15. A list of the questions represented in Figure 3 is provided in Figure 4, which can be copied and handed out for the students to answer.

Materials

1. At least one ant lion.

2. One 8-ounce (.25 l) yogurt cup or similar container per ant lion.

3. Sand to fill two-thirds of each cup.

4. Larger containers for Experiments 1 to 5.

5. Ants, fruit flies (see Chapter 16), or other small insects to feed the ant lions.

6. About 1 cup (.25 l) of sugar, salt, soil, coffee grounds, etc., per ant lion as additional building substances for Experiment 1.

7. Shelf paper that's sticky on one side for Experiment 6 (optional).

Background Information

Building the Pit

My husband once said that every time he goes into the study where I keep my thirty ant lions he hears the sound of sand flying. He's hearing sand grains hit the sides of sand-filled yogurt cups as the ant lions work to repair their pits, or prey traps. I love that sound. My ant lions are telling me that they're happy and healthy and busy doing the peculiar things they were born to do.

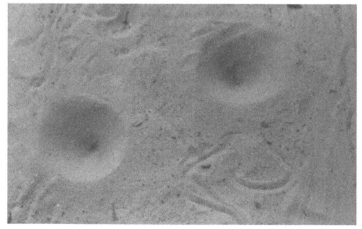

Fig. 12.1 Ant lion pits and trails.

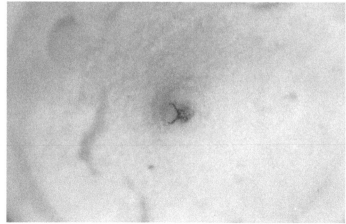

Fig. 12.2 Ant lion larva in sugar pit, with jaws open waiting for prey.

I try to imagine how a newcomer would perceive this scene. What meets the eye is a low-cut cardboard box holding thirty cups filled with sand and otherwise apparently empty. But if you follow the sound, you see in a few of the cups something tiny moving along just under the surface. A tiny head snaps back every few seconds, flinging sand. If you focus on one for as long as fifteen or twenty minutes you'll see the circular path of this mysterious creature spiral inward and downward as sand is tossed away. When the creature reaches the center of the spiral, the result is a conical pit (see Figure 12.1). The pit is smooth sided and tapers down to a perfect point. If your gaze wanders to the cups where there is no activity—those where the pit has been completed—you wonder what became of the architect. You have to look very carefully to see that it's still there. Its body is buried at the bottom point of the pit, hidden from its prey. But its jaws, its piercing caliper-like jaws, jut upward to receive its victims (see Figure 12.2). This is a unique kind of insect certainly; you can't relate it to any bug you've seen before.

Catching Prey

After you've watched an ant lion build its pit, the next thing to do is to feed it. Children love the drama. When you drop an ant or a fruit fly into the cup, the prey slides down the side of the pit toward the waiting jaws. If the victim tries to scramble up, the sand slides out from under its legs. When it reaches the bottom, you'll see the ant lion's jaws slam shut, piercing the prey between them (see Figure 12.3). Or you may just see a flurry of activity if sand dislodged by the prey's slide has obscured your view. The ant lion may slam the ant back and forth against the sides of the pit, trying to get a better grip with its jaws. If the ant lion misses on its

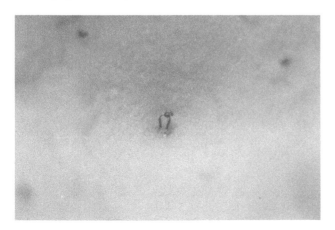

Fig.12.3 Ant lion larva in sugar pit, grasping fruitfly with jaws.

first grab, you may see sand grains flung precisely against the escaping prey, knocking it back down toward the waiting jaws again.

When the ant lion gets a good grip on the ant or other little bug, it draws the prey partially down into the sand and the jaws inject a paralyzing fluid into its body. The struggle soon stops. Then digestive juices, which dissolve the prey's internal organs, are injected through

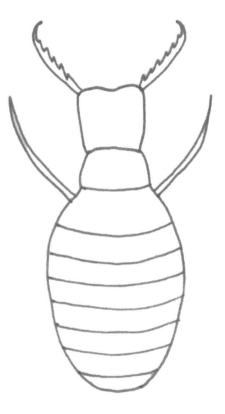

Fig. 12.4 Drawing of ant lion larva with jaws wide open. It has six legs but the back four are not usually visible from above. From $^1/_{16}$ to $^3/_8$ inch (1.6 mm–1 cm) in length, depending on age.

the jaws. The resulting fluid is sucked up through the jaws. (This is very similar to the way a spider eats, except that the spider uses a web instead of a pit to trap its prey. Both predators leave an empty husk of the prey behind.) The ant lion flips the empty carcass out of its pit with a quick flick of the back of the head. This head flick is its all-purpose manner of moving things—sand or finished prey. If you take an ant lion out of the sand, you'll see a plump oval body, a distinct flat head, and larger jaws, altogether no bigger than your little fingernail (see Figure 12.4). The largest are about $^3/_8$ inch (1 cm), the smallest $^1/_{16}$ to $^2/_{16}$ inch (1.6–3.2 mm). Their looks are not impressive, but their behavior is something else.

The Life Cycle

The ant lions that I keep for experiments are actually not mature insects, but larvae. They molt and shed their skins three times before they're ready to pupate. The larval period may last three years or more, depending on food supply. When an ant lion is ready to pupate, it spins a round cocoon about $1/3$ inch (.8 cm) in diameter. The cocoon is sticky and picks up sand grains (see Figure 12.5). I can tell when one of my ant lions has pupated because its pit will vanish or look in disrepair. If I start pouring the sand onto a newspaper I'll find the spherical cocoon close to the surface. I found four or five while looking for missing larvae before I realized that they weren't lumps of sand but sand-covered cocoons. The pupa stays in the cocoon for about a month, then the winged adult emerges. An adult is beautiful. It has a long skinny abdomen like a dragonfly and two pairs of long transparent wings (see Figure 12.5). The wings and body are both about 1 inch (2.5 cm) long. The adults are attracted to light and will fly to a window when they emerge, which is how I've caught the only living ones I've had. If you keep the pupa in an enclosure to capture the adult, check it every day because the adult will die if neglected. Some species eat fruit. I've kept one in captivity for several days, with two pieces of paper towel soaked with water and sugar water. After mating, the female lays her sand-grain-sized eggs in sand.

Fig. 12.5 *Ant lion pupa, larva, and adult.*

How to Get and Keep Ant Lion Larvae

I don't know of anywhere to order living ant lions, so you'll have to find your own. They are abundant in naturally sandy areas like sand dunes at the beach. There are very few naturally sandy areas where I live, but I am able to find ant lions pretty regularly in artificially sandy places like playgrounds. Many schools and city playgrounds now have sand. You'll find the ant lion pits only in places that are protected from trampling feet and heavy rain. For example, many wooden play structures have horizontal beams 2 to 12 inches (5–30.4 cm) above the ground. My children's school has a low balance beam, a wooden jungle gym, and a picnic table all in areas that have had sand added. They all have ant lion pits under them in the places where feet never go. I also occasionally find pits in wooded areas with very loose, fine soil.

An intact pit is easy to recognize. It's a very symmetrical cone-shaped hole in the sand, from $1/2$ to 3 inches (1.3–7.6 cm) across at the top and up to $1^1/2$ inches (3.8

cm) deep, depending on the age and size of the ant lion. A small pit looks something like the small hollows in the sand that form under an edge, from dripping water. But the pit has a point at the bottom, while a drip mark is more bowl-shaped. Once you've located a pit, you must have a spoon to get the ant lion out. Scoop down at least $1/2$ inch (1.3 cm) under the lowest part of the pit and bring up the sand. Then hold the spoon over a bowl or cup and gently shake the spoon back and forth, so the sand slowly falls over the spoon's edge, exposing the ant lion. Being somewhat buried in the spoon momentarily won't hurt it. Then put the ant lion in an empty cup to transport it. You can put them all in the same cup—in my experience they will not eat each other or anything else unless they're in their customary position at the bottom of a pit.

If you want the ant lion to live, it needs a pit. So set them up as soon as you get home. I keep my ant lions

Fig. 12.6 *Mandasa watching an ant lion burrow under the sand.*

separated at home, so they won't interfere with one another and there's no chance of cannibalism. Each ant lion needs a cup to itself, about two-thirds full of relatively fine sand, in order to make a pit. The sand must be dry and relatively free of lumps and leaf fragments. If you need to retrieve one from a cup of sand and it has not dug a pit, pour the cup of sand out slowly onto a piece of newspaper, spreading the sand out as you go. If you don't see the ant lion at first, gently shake the paper back and forth. Once you see it, you can scoop it up with a piece of paper.

Put the cups in a place where they won't be disturbed until it's time to take them to school. A bump may cause the sand to slide down the pit and cover the ant lion. Moving the cup almost certainly will.

After their pits are complete, feed them at least twice a week. Any tiny nonflying insect will do. I prefer wingless fruit flies because they're easier to keep and easier to control than ants. (See Chapter 16 for information on how to get and keep wingless fruit flies.) You can find ants by looking under logs or boards or putting some fruit out in your yard. I toss two or three apple cores in different parts of the yard, and I usually get ants on at least one. Ants are often found in loose soil, like that in gardens or flower beds, because it's easy to tunnel through. Use ants that are no more than half the length of the ant lion.

Field Hunt

If you take students on an ant lion field hunt, you may want to spoon them up yourself unless your kids can be very careful. Children can enjoy themselves just learning where to look for the pits and watching you catch the ant lions.

Ant Lions at School

Getting Ready

One ant lion per student is nice if you can find that many. I've used as few as one ant lion per class, though. Many of the experiments can be done with only one.

When it's time to go to school, I spoon each one out of its cup and put all the ant lions into one container, without sand. I take the cups to school empty, so the students can have the experience of setting up their own cups.

Observations and Activities

I start by explaining to the students that the ant lions we have are larvae, not adult insects. I draw a rough picture of the ant lion on the board, explaining that they are predators, describing the jaws and the pit and how they capture their prey. If I have enough ant lions, I tell them that everyone is going to get a cup of sand and an ant lion that they will put into the cup themselves.

After everyone has a cup of sand, I give each student an ant lion in the palm of his or her hand with a small amount of sand ($^1/_4$ teaspoon/1.25 ml or less). Very few students decline holding the ant lion, but those who do can receive it in the cup. As the student holds the ant

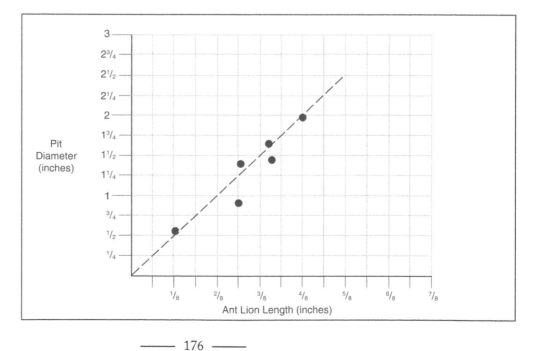

Figure 12.7 Sample graph showing a relationship between ant lion length and pit diameter. If the points lie along a straight or curved line (roughly), as these do, then there is a correlation between ant lion length and pit diameter.

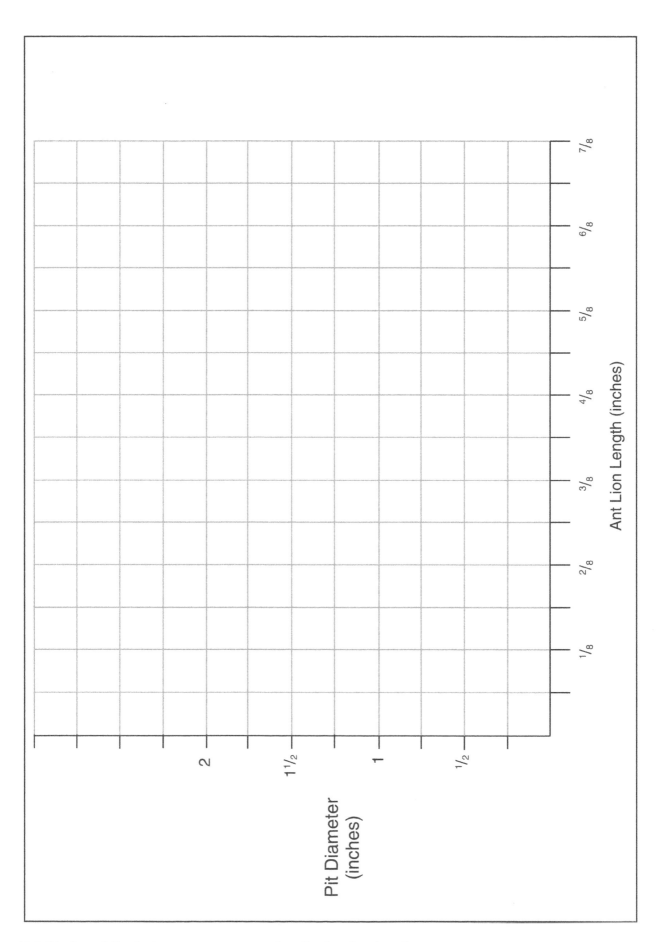

Figure 12.8 Blank graph for plotting your measurements of ant lion length and pit diameter.

Fig. 12.9 Stephanie shaking one fly from a fruitfly culture vial into an empty vial, preparing to feed her ant lion.

lion, point out the jaws, which will usually be wide open. (Ant lions never bite or try to bite people. Their jaws are for feeding only, not defense.)

The students will see the very fine tip on each jaw that pierces the victim. Closed jaws are harder to see. The ant lion will waggle its rear end up and down while walking backward trying to burrow under the sand. The small amount of sand in the student's hand allows the child to see how the waggling rear end works in the sand. Children find this delightful.

The first time I took ant lions to a class, I did all the handling myself, thinking the ant lions would get dropped or mashed. But I found with later classes that handling the ant lions was probably their favorite part, and not one was damaged.

At this stage, while the ant lions are uncovered, have each student measure his or her ant lion to the nearest millimeter. You can use this information later to see if there is a correlation between animal size and pit size. The class may also find a correlation between animal size and time to pupation.

I tell the students to put the ant lions in the cup after a minute or two of watching. Then they'll see the ant lion quickly bury itself, as it tried to in their hands (see Figure 12.6). In a few seconds it will be out of sight. One child asked, "How do ant lions protect themselves?" I answered his question with another question: "Why do ant lions bury themselves immediately?"

It would be nice to leave the ant lions on the students' desks all day so that they can watch the ant lions digging their new pits. But in order to see this, the students must be able to keep their hands completely

off the cup all day and refrain from bumping their desks at all. In my experience, most children can't do this, so I have them put their names on the cups and put them somewhere else.

The next morning most of the ant lions will have made new pits. Have each student measure the diameter of his or her pit at the top rim. You can make a graph for the whole class, plotting pit size against animal length (see Figure 12.7). Figure 12.8 provides a blank graph for your measurements. Is there a correlation? That is, do bigger ant lions make bigger pits? Yes—you will be able to see the relationship on the graph.

The next thing to do is feed the ant lions. I let each student put one prey insect, a fruit fly or an ant, into an empty fruit fly vial (or any small container) and feed his or her own ant lion. Even first graders can get a single fly out of a vial of reproducing fruit flies (for a culture vial, see Chapter 16). Tell them to knock the bottom of the culture vial hard on the palm of the other hand to shake the flies down from the top. Then remove the sponge from the top, shake one or two flies out into an empty vial and put the sponge plugs back in both vials (see Figure 12.9). Then invert the vial with one or two flies over the ant lion's pit and rap the bottom of it until the prey falls into the pit (see Figure 12.10).

Fig. 12.10 Sammy and classmates watch an ant lion in a yogurt cup grab at a fruit fly after Sammy dumps a fly from the vial into his ant lion's pit.

I find that I have to tell children constantly not to rap the container against the ant lion's cup, not to touch the cup in any way, and not to touch the sand. All these actions will cause sand to slide down the sides of the

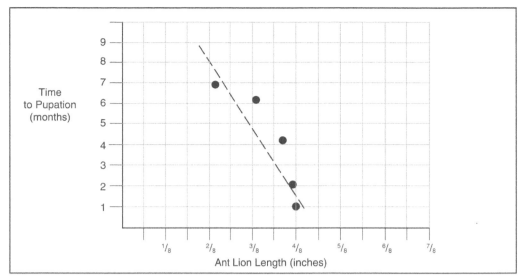

Figure 12.11 Sample graph showing a relationship between ant lion length and time to pupation. If the points lie along a straight or curved line (roughly), as these do, then there is a correlation between ant lion length and time to pupation. Your numbers may be different from these, depending on the animals' diet.

pit and cover the ant lion. Still, some do these things. I prefer to have students feed their ant lions in groups of three or four, at least the first time, so I can watch closely and make sure they understand about not touching the cup. After that they can do it on their own, usually first thing in the morning.

Pits in Sugar

A favorite activity of mine with ant lions is getting them to build pits in substances other than sand. Sugar works great. The white pits are beautiful, and you can see the ant lion's jaws so much better against the white background. You can also see ejected carcasses clearly around the rim of the pit. (They're very hard to see on sand.) Do the carcasses pile up all around the pit or only behind the ant lion? Have the students measure the distance between the ant lion and the fly and ant carcasses. Which carcasses are flung further? You can make another graph here, showing carcass size on the horizontal axis and distance from ant lion on the vertical axis.

Humidity causes a sort of crust to form on the sugar eventually, which helps would-be victims escape from the pit.

Spacing

If you have extra ant lions or not enough for everyone, put several together in a large box. I use one about 12 by 12 inches (30.4 x 30.4 cm), containing dry sand 2 to 3 inches (5–7.6 cm) deep. You can observe the spacing between the pits and the trails that the ant lions leave in the sand. Does spacing seem to be uniform (equal distances between neighboring pits) or random? What is the minimum distance between two pits?

Cannibalism

An interesting activity with ant lions in a box of sand is to try to push one ant lion into the pit of another. It's not easy to push one—they're well equipped for walking in sand and don't slide easily. After doing this, do the students think that ant lions often stumble into one another's pits while looking for new pit sites? What if you actually drop an ant lion into another pit from above? Does the resident ant lion kill the intruder? Sometimes. Can the intruder get out? Yes, if it escapes the jaws of the resident, it can easily burrow through the side of the pit to escape. Let the students discover this on their own.

Keep your ant lions until some of them pupate. Those that are $1/3$ to $1/2$ inch (.8–1.3 cm) are good candidates. All the pupations I've recorded have been in late spring and have followed a period of daily feeding. If you have body-length measurements for the organisms, you can create a graph that plots body length against the length of time to pupation (see Figure 12.11). Figure 12.12 provides a blank graph for your measurements. Larger ant lions can be expected to pupate sooner.

Experiments

Remember that the hypotheses I give are just examples. Your hypotheses will be the predictions made by the class or by a particular child. Your result for each experiment will be a statement of how your animals reacted to your experimental setup. Your conclusion is a statement of whether your prediction was confirmed or not confirmed. For each experiment, adding several replicates increases your confidence in the validity of your conclusion, but they may be omitted if tedious for young children.

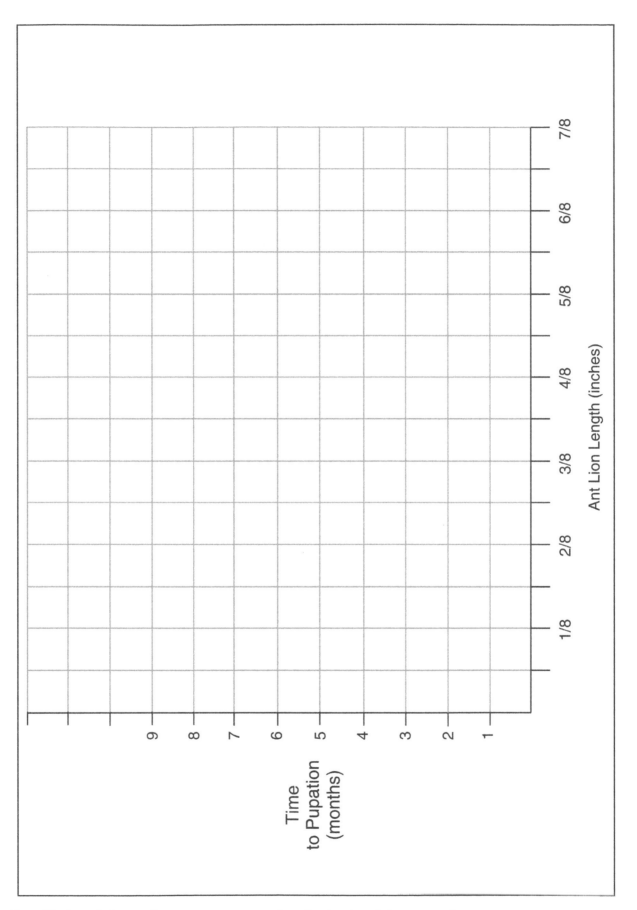

Figure 12.12 Blank graph for recording your measurements of ant lion length and time to pupation.

Building Material	Chosen for Pit Site (when offered another option)	Chosen for Pit Site (without another option)	No Pit Built
Sugar			
Salt			
Coffee Grounds			
Fine Soil			
Lumpy Soil			
Fine Sand			
Coarse Sand			
Any of the Above, Wet			

Figure 12.13 Blank graph for recording results for Experiment 1.

Experiment 1

Question:

If offered a choice between sand and some other building material, will the ant lion always choose sand?

Hypothesis:

Ant lions will always choose sand over other materials.

Methods:

Get a container at least 7 inches (17.8 cm) long, 4 inches (10.2 cm) wide, and 3 inches (7.6 cm) deep. The shape of the container is not important. Make a vertical partition dividing the container into two halves, each of which is big enough to accommodate a pit. Fill one side with fine dry sand up to the top of the partition. Fill the other side with sugar, coffee grounds, or soil—whatever material is being tested. Put an ant lion in the middle and record where it builds its pit. Repeat several times, if desired.

Result:

Your result is a description of the ant lions' choices. You can record the results on the table in Figure 12.13.

Conclusion:

Ant lions will build in many materials and don't always prefer sand. Texture seems more important than substance.

Experiment 2

Children hunting for ant lions on a playground will notice that pits are always under something. If they don't go with you on a field hunt, tell them you've noticed it. Why are the ant lions always under something? Is it because the ones who build in the open get stepped on, so we never see them? Or is it because they all choose to build under something? If they do choose to build under something, is it because they are attracted to darkness or dryness?

Question:

When given a choice, will ant lions build out in the open or under a cover (in shadow)?

Hypothesis:

If we give an ant lion a choice between cover and no cover, it will choose to build under cover.

Methods:

Put fine, dry sand at least 2 to 3 inches (5–7.6 cm) deep in a container at least 7 by 4 inches (17.8 x 10.2 cm) or big enough for two pits. Cover half of the container with opaque cardboard, about 1 inch (2.5 cm) or so above the surface of the sand. Leave the other side exposed. Put an ant lion in the center of the container and record where it builds its pit. Repeat several times, if desired. Do more build under cover or out in the open?

Result:

Your result is a description of the ant lions' choices. You can record the results on the table in Figure 12.14.

Conclusion:

In my experience, ant lions choose to build under cover somewhat more often than in the open, but the difference is not pronounced. Their prevalence under play structures can't be explained solely by a preference for darkness or cover.

Experiment 3

Question:

Do ant lions prefer dry sand to wet sand? (This could partly account for their prevalence under shelters.)

Hypothesis:

Ant lions like dry sand better than wet sand.

Methods:

Get a container that's at least the size described in Experiment 1 (a shoebox is fine). Add sand to a depth of 2–3 inches (5–7.6 cm). Put a foil or plastic partition in the sand in the middle of the box to divide the sand into two separate halves, then wet the sand on one side. The partition is to keep the dry side dry. It must not protrude above the surface of the sand. Put an ant lion in the center right over the partition, leave it a while, and record where it builds its pit. Repeat several times if desired.

Result:

Your result is a description of the ant lions' choices. You can record your results on the table in Figure 12.14.

Conclusion:

When offered a choice between wet sand and dry sand, ant lions build pits in dry sand. This preference for dry sand may partially account for their prevalence under structures. I often find them under narrow structures like balance beams, though, which probably offer little protection from rain. I also see them out in the open in protected areas. On the coast of North Carolina, walking is forbidden on some sand dunes in state parks to protect the sea oats that anchor the dunes. I can see from the beach that ant lion pits abound in these open areas protected from trampling. So it appears that human feet account at least in part for their absence in open areas of sandy playgrounds.

Ant lions often change the location of their pits. An ant lion traveling just under the surface of the sand leaves a shallow but easily traceable furrow in the sand. I've followed furrows through twists and turns for at least 10 feet (3 m) in remote sandy areas. Why do ant lions move their pits?

Experiment 4

Question:

Will rain cause the ant lion to change the location of its pit?

Hypothesis:

If we spray an ant lion with water regularly, it may move its pit.

Methods:

Put sand 2 to 3 inches (5–7.6 cm) deep in two boxes big enough to give the ant lions room to move around. I use boxes that are 12 by 18 inches (30.4 x 45.7 cm). Put at least one, preferably two or three, ant lions in each box. After they've dug their pits, thoroughly spray the pits in one box several times a day for a week or so. Don't spray those in the other box as a control (for comparison). Feed ant lions in both boxes normally. You can mark the location of each pit by sticking a toothpick in the sand next to it. Do more of the sprayed ant lions move their pits?

Result:

Your result is a description of the ant lions' movements during the week. Record your results on the table and histogram in Figure 12.15. Figure 7.4 provides an example of a completed histogram.

Conclusion:

In my experience, ant lions that are sprayed are more likely to move. Apparently, rain may cause ant lions to move their pits.

Experiment 5

Question:

Is a hungry ant lion more likely to change the location of its pit?

Hypothesis:

An ant lion that's not catching prey will move more than a well-fed one.

Methods:

Set up two boxes as in Experiment 4. Introduce one ant lion (two or three is better) into each box and allow it (or them) to dig pit(s). Feed the ant lions in one box daily for a week or two. Withhold food from those in the other box for the same period. Mark the location of each pit with a toothpick stuck in the sand beside it. Are the hungry ant lions more likely to move their pits?

Result:

Your result is a description of the ant lions' movements during the week. Record your results on the table and histogram in Figure 12.16. Figure 7.4 provides an example of a completed histogram.

Conclusion:

In my experience, a hungry ant lion is more likely to move than a well-fed one.

Experiment 6

Question:

How far do ant lions fling sand, and how far do they fling sugar when making pits? Are the distances the same?

Hypothesis:

Ant lions fling sand and sugar the same distance.

Methods:

Put shelf paper down on a table, sticky side up so the sand and sugar will stick to it and not slide. Put one ant lion in sugar and one in sand 2 to 3 inches (5–7.6 cm) deep in cups on the shelf paper. (Shelf paper is not essential if the cups are in an

Choice of Conditions Offered	Ant Lion's Choice	Effect of Choice on Ant Lion*
Under Cover or Out in the Open Experiment 2		
Dry Sand or Wet Sand Experiment 3		*Cover protects from feet and rain *Wet sand is not flingable

Figure 12.14 Blank table for recording results for Experiments 2 and 3.

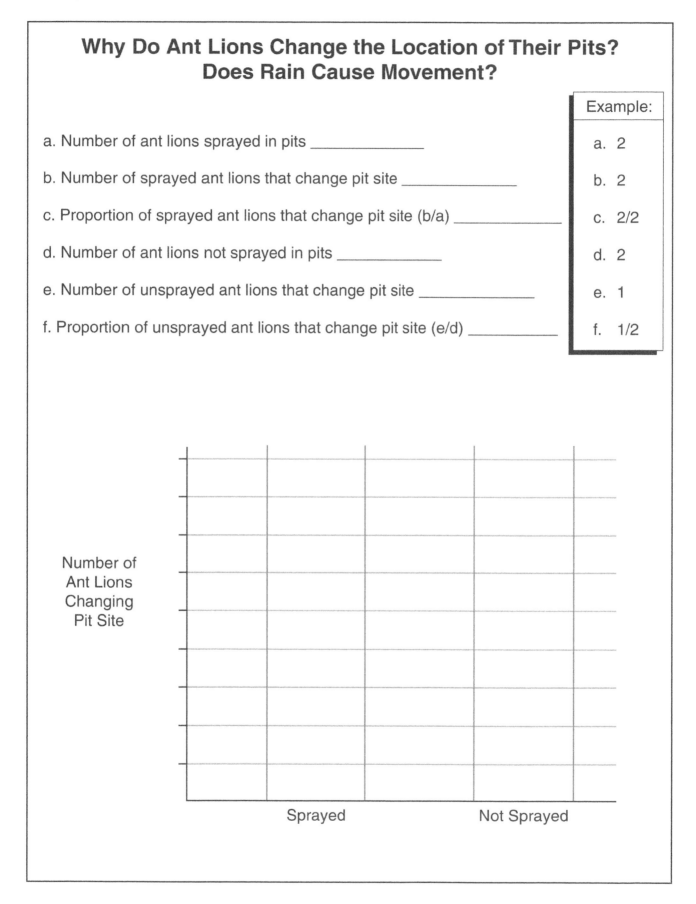

Why Do Ant Lions Change the Location of Their Pits? Does Rain Cause Movement?

a. Number of ant lions sprayed in pits _____

b. Number of sprayed ant lions that change pit site _____

c. Proportion of sprayed ant lions that change pit site (b/a) _____

d. Number of ant lions not sprayed in pits _____

e. Number of unsprayed ant lions that change pit site _____

f. Proportion of unsprayed ant lions that change pit site (e/d) _____

Example:

a. 2

b. 2

c. 2/2

d. 2

e. 1

f. 1/2

Number of Ant Lions Changing Pit Site

Sprayed   Not Sprayed

Figure 12.15 Table and blank histogram for recording the number of sprayed ant lions that move pit site versus the number of unsprayed ant lions that move pit site. (See Figure 7.4 for an example of a completed histogram.)

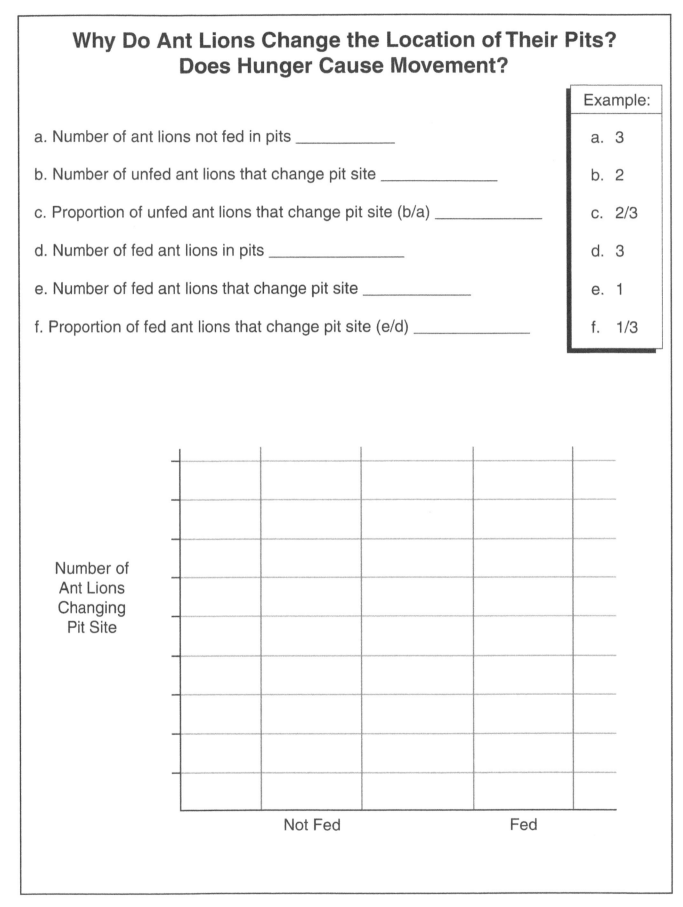

Why Do Ant Lions Change the Location of Their Pits?
Does Hunger Cause Movement?

	Example:
a. Number of ant lions not fed in pits _____	a. 3
b. Number of unfed ant lions that change pit site _____	b. 2
c. Proportion of unfed ant lions that change pit site (b/a) _____	c. 2/3
d. Number of fed ant lions in pits _____	d. 3
e. Number of fed ant lions that change pit site _____	e. 1
f. Proportion of fed ant lions that change pit site (e/d) _____	f. 1/3

Number of
Ant Lions
Changing
Pit Site

Not Fed   Fed

Figure 12.16 Table and blank histogram for recording the number of well-fed ant lions that move pit site versus the number of unfed ant lions that move pit site. (See Figure 7.4 for an example of a completed histogram.)

area that won't be disturbed or swept.) Give them a day to make pits, then measure the diameter of the circle of scattered sugar and sand around the respective cups.

Result:

Your result will be a description of exactly what happened and a statement of your distance measurements. Record your results on the table and histogram in Figure 12.17. Figure 7.4 provides an example of a completed histogram.

Conclusion:

Was your prediction confirmed or not? In my experience, ant lions fling sugar farther than sand. Ant lions on my desk in cups with sugar fling sugar all over my papers, but the ones in sand are much neater. Sugar must be lighter than sand. You can confirm this by weighing equal volumes of sugar and sand in equal containers.

Measure the Diameter of the "Circle" of Scattered Sugar and Scattered Sand around Your Cups.

If you had several cups of each type of substrate, calculate an average diameter for each type.

a. Diameter of circle of scattered sugar _____
(single measurement or average)

b. Diameter of circle of scattered sand _____
(single measurement or average)

Which substance was flung the farthest? _____

Was your prediction confirmed? _____

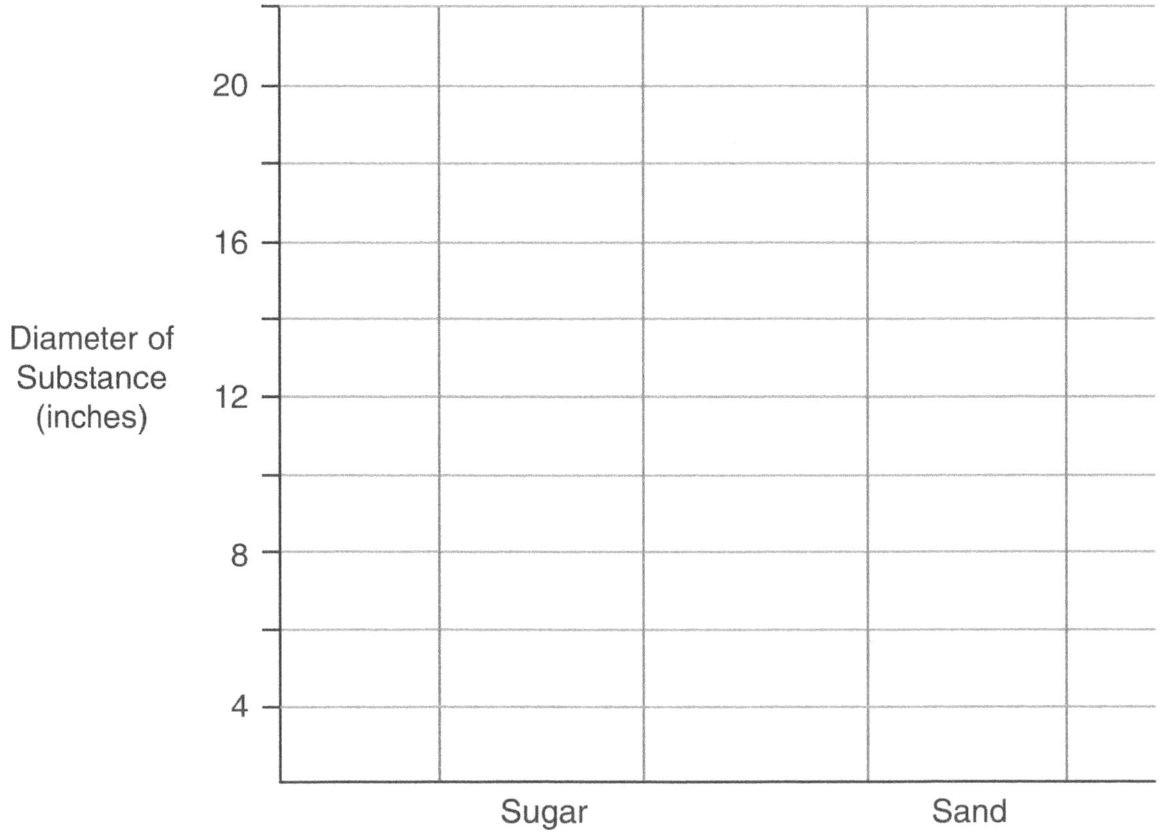

Diameter of Substance (inches)

20
16
12
8
4

Sugar   Sand

Figure 12.17 Blank histogram for recording results for Experiment 6. (See Figure 7.4 for an example of a completed histogram.)

Chapter 13
Spooky Spiders

Introduction

In this chapter, students will discover through experimentation that some spiders need webs to catch their prey but some do not. They'll see how inflexible each type of spider is in its method of prey recognition and capture. Each type of spider must have a specific signal, or stimulus, to recognize an insect as a meal. Without the signal the spider won't attack. Most children are unable to predict the outcome of the experiments, and the surprise element makes a big impression.

Figure 3 in the Postscript provides a range of questions and answers comparing behavior of the predators discussed in Chapters 12 to 15. A list of the questions represented in Figure 3 is provided in Figure 4, which can be photocopied and handed out for the students to answer.

Materials

1. Funnel weaver, orb weaver, cobweb weaver, or other web-building spider.

2. Jumping spider or nursery web spider or wolf spider or other "wandering" (non-web-building) spider.

3. Plastic jar or other enclosure for each spider. I often use plastic peanut butter jars, approximately 1 quart (1 l) size.

4. Lids with air holes (or cloth lids and rubber bands) for the jars.

5. Plastic petri dishes as alternative enclosures for jumpers and other small non-web-builders. These allow excellent visibility. Petri dishes are available at science hobby shops or biological supply companies (see Appendix).

6. Fruit fly culture vials are also good alternative enclosures for small non-web-builders. (See Chapter 16 for information on how to get vials.)

7. A terrarium for Experiments 4 and 5.

8. Prey for spiders (suggestions below).

Background Information

Categorizing Spiders

There are thousands of species of spiders. There is no need to identify any of them to species, and I don't try. It's easy to recognize a few broad categories of spiders. I use an inexpensive paperback from the Golden Field Guide series, called *Spiders and Their Kin*. The most obvious major categories of spiders are these: web-building spiders and wandering spiders. Web-builders rely on their webs for prey capture and usually stay on the webs at all times, which makes them easy to recognize. I use the term "wandering spiders" to mean any spiders that don't build webs for prey capture, but either ambush prey or actively pursue it. (There is a family of tropical spiders that are sometimes called "wandering spiders," not to be confused with my much broader use of the term.) The wanderers can be recognized by the fact that they're not on webs but rather hiding or walking across a sofa or a sidewalk or through the grass, and so forth.

Children generally think that all spiders build webs and that they all build the circular webs with spokes that we see in storybooks. But the circular web with spokes, or orb web, is only one of many different styles of web. There are platform webs, cobwebs, funnel webs, and so on. Closely related spiders build similar webs, so an easy way to categorize web-builders is by type of web: cobweb weavers, funnel weavers, orb weavers, and so forth. There are only a few common web types, and they're easy to

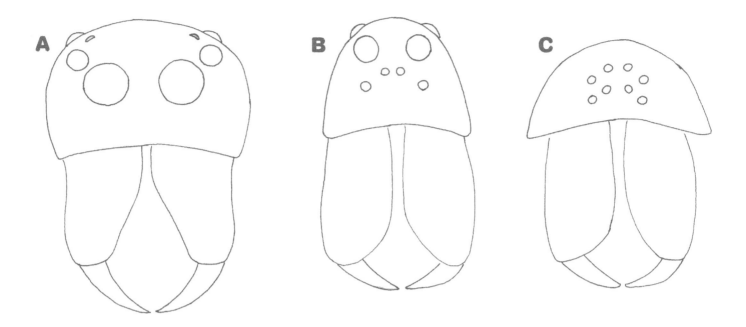

Fig. 13.1 Three spider faces. Compare eye size of (a) jumping spider, (b) wolf spider, and (c) funnel weaver. Also note fangs on tip of jaws.

recognize. *Spiders and Their Kin* helps by having spiders grouped by web type. The non-web-builders are grouped according to behavior or body shape, and the most common groups are not hard to recognize or remember. Jumping spiders jump and have short, strong, stocky legs; crab spiders have legs that project sideways and then curve forward like crabs' legs; wolf spiders are big and hairy, and so on.

Spider Anatomy and Basic Behavior

All spiders have eight legs and two main body parts: a cephalothorax and an abdomen. This distinguishes them from insects, which have six legs and three main body parts: head, thorax, and abdomen. The cephalothorax of the spider is equivalent to the head and thorax of the insect. Like insects, spiders molt or shed their exoskeletons several times as they grow to adulthood.

Most spiders have eight eyes (see Figure 13.1). The size of the eyes is related to the quality of vision. Jumping spiders have much better vision than other spiders. They have two eyes in front that are much bigger than the other six and can form images as well as detect motion. Jumpers are active hunters and rely on their acute vision to detect prey. Many other wandering spiders, like the wolf spider, rely on vision to detect prey, but can detect motion only. For a sit-and-wait predator like the wolf spider, this type of vision is adequate. Most web-builders, like the funnel weaver, have very small eyes

and poor vision. They generally rely on vibrations in the web to locate prey.

Because all spiders are predators (which means they eat live prey), they almost all have the same body parts for holding and eating prey. Each spider has a pair of mouth appendages called pedipalps that grasp the prey. Some pedipalps are quite large. (The pedipalps on a trap-door spider are so large that on my first encounter I mistook them for a fifth pair of legs.) Between the pedipalps,

Fig. 13.2 Jumping spider. The front legs are held up as in courtship, to display the hairy ornamentations.

on the front of the head, are the jaws or chelicerae (see Figure 13.2). At the tip of each of the two jaws are the fangs. The spider injects poison into its prey through the hollow fangs. After the poison has killed or paralyzed the prey, the spider injects digestive juices, which liquefy the insides of the prey. (Some spiders spit digestive juices.) The spider then sucks up the insides and leaves a hollow shell of the insect. Some spiders have spines on the inner edges of the jaws and crunch the hollow body into a wad.

All spiders secrete silk from a cluster of glands on the tip of the abdomen, called spinnerets (see Figure 13.3). The non-web-builders use the silk only for wrapping up prey or making egg sacs or as a safety line (dragline) when moving around. I poked a webless spider under the eave of my house one morning, and it dropped halfway to the ground on a dragline in half a second by affixing some silk to the eave and secreting a line as it dropped. I was amazed at how fast it secreted that line when it perceived itself to be in danger. It jerked to a stop right in front of my face.

Spiders are not my favorite creatures to handle, but their behavior is interesting. There's a lot of action with a captive spider. Some predators won't eat out in the open, like the centipedes I've kept, but spiders don't care who's watching. My four-year-old, Alan, has a "pet" nursery web spider (very similar to a wolf spider but less hairy) that he found in the living room and caught with a plastic peanut butter jar by himself. It's been in the jar for about a month. He feeds it a housefly every other day or so, which he catches bare handed against the living room windows. His spider is about $1^1/_2$ inches (3.8 cm) long, including the legs. He unscrews the lid, throws the usually wounded fly in, then puts the lid back on and watches. The spider is on the fly as soon as it moves. Nursery web spiders spit digestive juices on their prey and crunch its outer skeleton as they extract the insides. When the spider is through, the fly is just a tiny wad.

Alan has a pet jumping spider too that he caught outside on a water fountain at the park. It ate small fruit flies at first, but has now grown and molted and can handle Alan's houseflies. It doesn't crunch its prey but merely sucks them dry, so its jar is littered with fly carcasses. Neither of these two are web-builders.

The first jumping spider we kept captive at home made herself a silken enclosure about two weeks after we got her. She sealed herself in and wouldn't come out. We left her alone on a shelf in our study. Several weeks later my husband found some baby spiderlings on his desk. It finally occurred to us to check the silken enclosure. That was the source—dozens were coming out. We tore it

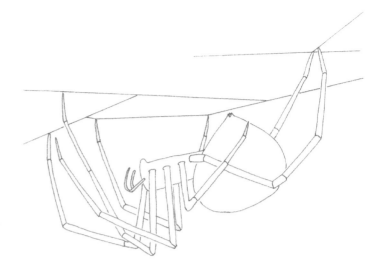

Fig. 13.3 American house spider in usual upside-down posture. Points on tip of abdomen are spinnerets.

open and found the mother's shriveled body. If this happens to you in a class, reading E. B. White's *Charlotte's Web* may help the students savor it. Charlotte the spider dies after making an egg sac, and the hatchlings are later greatly appreciated by Charlotte's pals.

How to Get and Keep Spiders

I don't know of a place to order specific types of spiders other than orb weavers and tarantulas, and they are rather expensive (see Appendix). But spiders are so common that it's easy to catch your own. To do the experiments I've described here, you need one or two web-building spiders and one or two non-web-builders (wanderers).

How to Catch a Web-Builder

I use funnel weavers for the experiments that require web-builders because they're so abundant around my house. Funnel weavers hide at the narrow end of a funnel-shaped web (see Figure 13.4). The floor of the web spreads out into a sheet. There are probably one hundred funnel webs in the ivy across the front of my small house. Many are in the vertical crevices between the downspouts and the wall, uniformly spaced at intervals of about 7 inches (17.8 cm). The symmetry reminds me of a high-rise apartment. On an ivy-covered tree behind our house, there are only four webs, which are spaced at wider but just as regular intervals.

To catch funnel weavers and most spiders, slap a plastic jar down over the spider as fast as you can. After you have the jar over the spider, slide a piece of cardboard between the jar and the wall or ground. Then pick up the jar and replace the cardboard with a lid. Funnel weavers will build a modified web in a jar, which often winds up looking more or less like a silk doughnut lying on the bottom of the jar. The doughnut is a circular tunnel where the spider resides, with irregular strands of silk filling the space above the tunnel. The spider emerges from the tunnel when a victim is snagged on the strands above it.

Fig. 13.4 *The web of a funnel weaver spider, one of hundreds in the ivy on the front of my house.*

If you have a jar 1 quart (1 l) or larger, you don't have to have holes in the lid as long as you lift the lid for a moment every day. (This may disrupt the web, so you may prefer air holes.) Feed the spider small insects like flies or moths or young crickets twice a week or so.

If you can't find a funnel weaver, any web-building spider will do. American house spiders (this is their name) are very common. They make tangled, irregularly shaped webs that are seen in corners, especially ceiling corners, in houses. This type of web is called a cobweb. The house spider has a round abdomen that's huge compared to the rest of its body, very frail-looking legs, is light brown, and has black streaks and patches (see Figure 13.3). The twenty or so that are behind curtains and windows, behind toilets, and on bedroom ceilings in my house right now are $^3/_4$ inch (1.9 cm) or less, including legs. House spiders are easy to catch with a jar, and they build cobwebs readily in the jar.

Black widows are also cobweb weavers and have a body shape that is identical to that of the American house spider. (See the Field Hunt section for more on

black widows.) House spiders don't have a dangerous bite and probably don't bite at all.

Most web-building spiders will build some sort of web in a jar, but may not if their normal web is a large orb web or some other type that can't be even approximated in a jar.

How to Catch a Wandering Spider

Jumping spiders are my favorite and are easy to catch too. I stumble upon one every couple of weeks or so. They're constantly on the move, looking for prey. I find them in odd places—my car, the kitchen table, the living room floor. Most are easy to recognize because of their short stocky legs (see Figure 13.5). They walk at an irregular pace and may jump several centimeters when going after prey. Most jumping spiders are quite small, up to $^3/_8$ inch (1 cm), excluding the legs. A *National Geographic* article on jumping spiders called them "teddy bearish," and they are. Many are fuzzy or have pretty tufts of bristles or bright colors to attract mates.

Fig. 13.5 *Jumping spider. Note stocky legs, large eyes in front center.*

All jumping spiders have a pair of very big eyes (by spider standards) on the front of the head (cephalothorax), which gives them a sentient look. They also have six smaller eyes that are much less noticeable. These spiders have excellent vision and will turn to face anything moving around them. The big pair of eyes on the front of the face, along with the turning behavior, can help you recognize them as jumping spiders.

The best way I know to catch a jumping spider predictably is to look around the top of fence posts. They like high places—walls, tree branches, and so on. There's a jumper at the top of almost every single post of a chain-link fence around a baseball field behind my house in late summer and fall. They hide in silken cocoon-like envelopes, open at both ends. If I poke a twig in one end, the spider runs out the other end into the jar (a second person to hold the jar helps).

If you can't find a jumping spider, any wandering spider (non-web-builder) will do. Look carefully in the grass for a nursery web spider (see Figure 13.6). I find

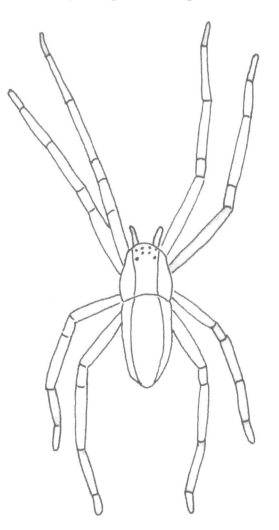

Fig. 13.6 Nursery web spider. Very similar to wolf spider, but less hairy.

them most predictably in overgrown grass that's been blown or pushed down to make lots of caves underneath. Wolf spiders are also found in such places and look very similar to nursery web spiders. Both are large, ³/₄ inch (1.9 cm) in length, excluding the legs. In the fall, females are likely to have egg sacs. A female wolf spider will have her egg sac attached to her spinnerets. A female nursery web spider will carry her egg sac with her jaws until near hatching time, then she will hide it in a folded leaf.

Any spider you find in the grass or on the forest floor that's not in a web is likely to be a wandering spider. If you look, you will find one! Catch it with a jar and small square of cardboard. Wandering spiders may make draglines crisscrossing the inside of a jar, which can be mistaken for a web. If in doubt, put the spider in a terrarium where it has room to get away from its draglines and make a proper web if it's so inclined.

Housing and Feeding Your Spiders

I usually keep spiders in totally empty plastic peanut butter jars. There is no reason to add soil or leaves to the jar; it will only make web building more difficult and obscure the prey. I keep small jumping spiders in plastic petri dishes because I can see their movements so much better in a petri dish than in a jar, especially their reaction to prey and to each other. Fruit fly vials work nicely too for small non-web-builders. (Petri dishes and fruit fly vials are described in Chapters 11 and 16, respectively.) Large web-builders may need a terrarium.

If spiders are fed adequately, they don't need water. I feed them a couple of times a week. Sweep a net through the grass. Almost any little insect out of the grass will do, although it has to be moving. Flies are favorites. Moths are also good prey. You can catch a moth with a porch light at night and a net. Insect larvae (caterpillars, beetle grubs, maggots), other spiders, and small bees may or may not be accepted. Pill bugs (roly-polies), slugs, millipedes, and worms will probably not be accepted. The prey should not be so large that it completely wrecks the spider's web and should be smaller than the spider itself.

Field Hunt

If the students want to catch their own spiders they can look through leaf litter, in the grass, in shrubs, around windows, on outdoor walls, and atop fence posts. Record whether each captive was in a web or not. Substitute your captives for the ones I've suggested for each category: web-builders and wandering spiders.

Poisonous Spiders

Don't let a fear of bites keep you and your students away from spiders. Most spiders either won't bite because they are too timid or can't bite because they're too small. It's still probably a good idea to tell students not to handle any spider. The black widow and the brown recluse are the only two whose bites are dangerous. The bite of any other spider in the United States is either painless or equivalent to an insect bite or sting.

Black widows are virtually identical in shape to the American house spider (see drawing) but shiny black all over except for a red hourglass marking on the underside of the abdomen. I live in the Southeast, where black widows are most common. Most of the black widows I've found have been in dark places like basements or crawl spaces under houses or dark gas station restrooms or dark cabinets inside my house. The only place I've ever seen them outdoors is a couple of times under big heavy logs or heavy stones. I've never seen one out in the open outdoors. In many parts of the country, especially the northern states, you'll never see black widows at all.

The brown recluse is found in the same type of setting as the black widow—dark corners and out-of-the-way dark cabinets indoors and under heavy stones outdoors. It is yellowish brown and has a dark violin shape on the front half of the cephalothorax. I've never seen one.

Both the black widow and the brown recluse can be avoided by telling students to keep hands out of dark places that they can't see into.

Spiders at School

Getting Ready

All you need to begin observing is a plastic jar with a lid and prey for each spider. Halloween is a good time to do a unit on spiders if you haven't had a hard freeze by then, which would kill many of the adult spiders outdoors. The big orb weavers, like the conspicuous black and yellow *Argiope*, are most easily found in fall. This is the time when many other adult spiders are most abundant after growing and molting all summer.

Observations and Activities

Web-Builders

Some children may not know why spiders have webs. Put a funnel weaver or whatever web-building spider you can find in a jar or terrarium and give it a day or two to build a web. Then throw in a fly or lightning bug or other insect and watch. When the prey gets caught in the web, the spider will grab it and probably wrap it up in silk to hold it securely. The spider will bite the prey to paralyze it and may eat it then (suck it dry) or wrap it more thoroughly and save it for later. If the spider ignores it, the prey may be too big or not moving enough, or the spider may be too concerned about its own safety to be hungry.

Watch then for the children's curiosity to take over. Can we trick the spider by jiggling a twig held

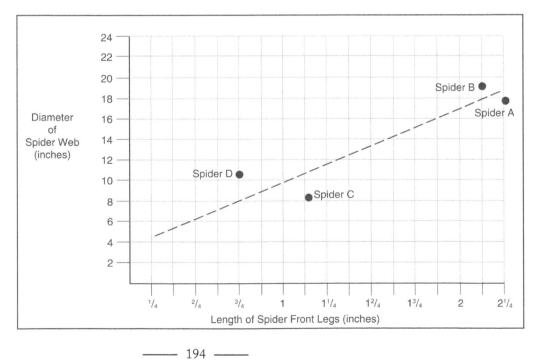

Figure 13.7 Sample graph showing a relationship between spider leg length and web diameter. The points will probably not conform exactly to a line, but will more or less if your spiders are all the same species.

against some part of the web? Does the spider attack? Sometimes. What does the spider do with an empty carcass?

Do bigger spiders have bigger webs? Yes, there is a direct relationship between the length of the spiders' legs and the size of the webs. If you have several web-builders of the same type, different sizes, you can show this on a graph (see Figures 13.7 and 13.8). Let the horizontal axis be spider-leg length and the vertical axis be the web diameter. If you plot several points for several spiders and their respective webs, you should get more or less a straight line or possibly a curved line. You can use that line to predict web diameter for any new spider of the same type. It's easiest to do this with orb weavers, since you don't have to tear up the web to measure the spider. You can record your measurements on a graph (see Figure 13.7). Figure 13.8 provides a blank graph for your measurements.

Wandering Spiders

If you find a spider with an egg sac attached to her abdomen, which some mother spiders do to protect the eggs, remove the egg sac and offer a substitute—a bit of paper or other lightweight object about the size of the egg sac. What does the mother spider do? A mother wolf spider may attach the substitute object in place of the egg sac.

How does a wandering spider react to prey? Most will either ignore the prey or approach rapidly and attack it. They see something small moving and they go for it. Many wandering spiders are able to detect motion only. Jumping spiders are different, though. The jumper's two big front eyes have good resolution—that is, they are able to form a clear image of the prey. A jumper's behavior is like that of a toad or a mantis when it first notices the movement of the prey. It will turn the front part of its body to face the prey directly, to fix those two big front eyes on it.

This is nice because it signals the viewer (you or me or a student) that the predator has definitely noticed the prey. One advantage of the petri dish for a jumper is that you can look down on the spider and so can see its body swivel more clearly. Can the students tell when the jumper has noticed a new prey item? Get them to articulate how they can tell.

After the jumper has noticed the prey, it may watch for a few moments before jumping on it, or it may decide not to attack. It will almost always turn to face directly something new that's moving, at least momentarily. We

have a couple of jumping spiders in petri dishes on the kitchen table where I work. They are startled and hop backward or at least watch me every time I pick up a book or a paper. In contrast, the spiders in webs in jars never react to me or my activities at all unless I touch the web or pick up the jar. They can't see me, but the jumpers can.

It's fun too to watch a pair of jumping spiders react visually to one another. I have a same-species pair together in a jar now. (I believe they are the same species because they have identical markings.) The smaller one (not mature) reacts to the larger one by turning to face it head-on whenever the larger one moves, and by walking backward to get away from it. The smaller one may turn and run if the big one comes too close. The larger one has taken over the smaller one's silken hideout and now keeps the smaller one out. The larger one also gets most of the prey I put in the jar and may eat the smaller one eventually.

If you have two adult jumping spiders of the same species you may see some courtship displays or male-to-male aggression. Adult males usually have some flashy colors or tufts of hair on the front legs and maybe elsewhere. Female jumping spiders, like female birds, are more usually drab brown or gray. During courtship, males lift their front legs to display their colors and tufts to the females (see Figure 13.2). Two males may display to one another competitively. They may also butt heads or otherwise show aggression. A jumping spider will react to its mirror image as though it's another jumper. A male may turn to face its image and behave aggressively toward it.

Will a wolf spider or nursery web spider or other wandering spider react to its mirror image? Probably not. Although they do detect prey visually, many only detect motion and can't form images at all. A wolf spider can't see its mirror image because it sees only motion. When the wolf spider is walking, everything is a moving blur. When it remains still, its mirror image is still too. Wolf spiders are ambush or sit-and-wait predators, so this type of vision is adequate for them in nature. The spider usually waits motionless, and only the prey is moving, so the prey is visible as movement.

Put a pair of jumping spiders together, a pair of wandering spiders other than jumpers together, and a pair of web-builders together. Have the students compare how the pairs react to each other. Jumpers see the best, other wanderers see somewhat, and web-builders usually see poorly. Do the students' observations confirm this?

Experiments

Remember that the hypotheses I give are just examples. Your hypotheses will be the predictions made by the class or a particular student. Your result for each experiment will be a statement of how your animals reacted to your experimental setup. Your conclusion is a statement of whether your prediction is confirmed or not. For each experiment, adding replicates increases your confidence in the truth of your conclusion, but they may be omitted if tedious for young children.

Experiment 1

Here's a question one child posed after several days of watching a funnel weaver spider eat prey in its web in a jar.

Question:

If we put another spider in the jar, would it get caught in the silk like a fly? Or could it walk across the silk like the one that lives in the jar?

Hypothesis:

If we put in a second spider, we think the new spider can walk on the first spider's silk without getting caught.

Methods:

This is an easy question to test. Allow one spider to build a web in a jar, and add a second spider, either of the same or a different species. Watch and record what happens. Your control here is watching the intruding spider (or another identical one) on its own silk first.

Result:

Your result will be a description of what happened between the intruding spider and the resident spider.

Conclusion:

I've had different outcomes here. Often an intruder is able to walk on the existing web, but the resident usually kills it, eventually.

What if the intruder is much bigger than the resident? Does the intruder ever kill the resident or coexist with it? If it did kill the resident, what would it do about the resident's web? Use it? Repair it? Spiders often eat their own damaged webs before rebuilding. Would a victorious intruder eat the resident's web to make room for its own web?

Does the result vary according to whether you use a same-species pair of spiders or two different species?

Experiment 2

Question:

Does a funnel weaver or other web-building spider have to have its web to catch prey?

Hypothesis:

If we destroy the funnel weaver's web or put the funnel weaver in another jar, it will still eat prey offered to it.

Methods:

Put a funnel weaver or other web-builder in a fresh jar. As a control, keep another funnel weaver in a jar with a web. Offer both prey, such as a small, immature cricket. Do both eat?

Result:

Your result is a statement of how both spiders responded to the prey.

Conclusion:

A funnel weaver in a jar with a web will eat prey readily. A web-builder of any sort will not (in my experience) eat prey without a web. The prediction is not confirmed.

Web-builders are genetically "programmed" to recognize prey by the vibrations of the web. Their poor eyesight is not helpful; the vibrations alone trigger the prey-capture response.

Experiment 3

Question:

Do all spiders ignore prey if they are not in a web?

Hypothesis:

After seeing the funnel weaver in Experiment 2, we think all spiders have to have a web to catch and eat prey.

Methods:

Put a funnel weaver or other web-builder in a fresh jar. Put a jumping spider or other wandering

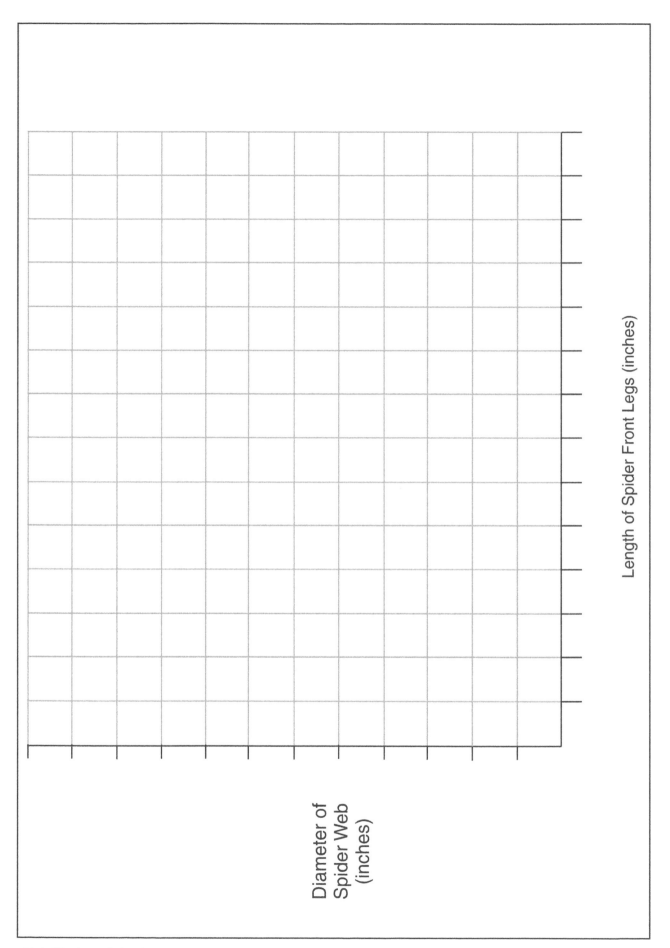

Figure 13.8 Blank graph for plotting measurements of spider leg length versus web diameter.

spider in another fresh jar. Larger ones may require a small terrarium. The spiders must be in the jar long enough to get calm (an hour or so) but not long enough to build draglines all over the jar (even non-web-builders may leave draglines when moving around). Put a fly, small cricket, or other small prey in each jar. Does either eat?

Result:

Your result is a statement of the response of both spiders.

Conclusion:

The prediction here is, of course, not confirmed. Having no way of detecting the prey without a web, the web-builder will ignore the prey. But the wandering spider should jump on the prey or otherwise attack it, and either eat it or wrap it up for future use. This confirms that most wandering spiders depend on eyesight and not vibration to detect prey. (If the jar has been passed around recently, the spider may be too disturbed to eat.)

Experiment 4

Question:

Do spiders have preferences about where they build their webs?

Hypothesis:

Whatever the students come up with.

Methods:

Set this up as a choice experiment in a terrarium big enough so that either end of it could accommodate a web. Offer different conditions at either end; for example, light and dark, damp sand and dry, a round ball on one end of the terrarium floor versus open space at the other end, an angular object to maximize possible attachment points (like a small, lidless box) in one end of the terrarium versus open space at the other end, a dish of water in one end versus an empty dish, and so on. Since for any trial the spider could select one end out of indifference instead of a strong preference, you'd have to repeat it several times to be confident of a preference.

When a web is damaged, many spiders will eat the whole web (which conserves energy) and start over. So after each choice, you can damage the web and

leave it in place to see if your spider does eat it and start over.

If you have a non-web-builder, you can offer it choices too and see where it spends most of its time. If it's a sit-and-wait hunter, where does it hide?

Result:

Your result will be a statement of your observations of your spiders.

Conclusion:

Depends on the type and size of the spider and the choices offered.

Experiment 5

Question:

Will tearing the web up daily or spraying the spider with water cause the spider to change the location of its web? Can the spider be induced by other means to change the location of the web?

Hypothesis:

We think the spider will build in a new location if we damage its web.

Methods:

Damage the web somewhat each day with a stick or spray the spider with water or disturb it in some other way. Withholding food from it is one option. Will hunger make it move? The spider has to be in a terrarium or other container big enough to have another choice of locations. Your control is another spider of the same type that is not being bothered, sprayed, or starved.

Result:

Your result is a statement of your spider's reaction.

Conclusion:

Some spiders can be remarkably persistent about building a new web daily in the same place. After all, webs are torn up a lot in nature. Some spiders routinely eat the whole web every night and start over. Others may move if disturbed, particularly if they themselves are menaced. Water may or may not bother the spider, depending on the species and the force of the water. Usually not. In my experience, starvation does not usually cause a spider to move.

Experiment 6

Question:

If two funnel weavers are in the same jar and there's not room for two webs, will they build a web together?

Hypothesis:

We think two crowded spiders will build a web together.

Methods:

Put two funnel weavers or other web-building spiders in a jar together. As a control, have a single one in another jar to make sure this type of spider will build in a jar.

Result:

Your result is a description of the actions of your spiders.

Conclusion:

In general the prediction here is not confirmed. Usually neither will build a web. One may, or possibly both may squeeze in a small web, but certainly they won't build one together.

Experiment 7

Question:

Will spiders build webs in the dark?

Hypothesis:

If we put the jar in the dark, the spider can't see to build.

Methods:

Put two web-building spiders of the same type in two jars. Put one jar in the dark and leave one in a well-lit room. Does the one in the dark build a web as well as the one in the light?

Result:

Your result is a statement of your spiders' activities.

Conclusion:

The prediction here is not confirmed. Since web-builders have poor eyesight, they don't rely on vision much or at all for web building. Most can build a normal web in the dark. Can a web-builder catch prey in the dark?

Experiment 8

Question:

How does darkness affect prey capture?

Hypothesis:

Since web-builders can't see well, we expect that they can catch prey in the dark. Since wanderers, especially jumpers, see well and use their vision to catch prey, we expect that they can't catch prey in the dark.

Methods:

Have a web-builder with web intact in one jar. In another jar have a wandering spider, preferably a jumping spider. Take both jars into a closet or other room that can be made very dark. Add prey, such as a small cricket or moth, to each jar at the same time and turn off the light quickly. Check the spiders after ten minutes or so. Have both caught their prey?

Result:

Your result is a statement of what happened with your two spiders.

Conclusion:

The prediction is confirmed. Since web-builders rely on tactile detection of web vibrations to find their prey, darkness has little effect on their prey capture. Since wanderers rely mostly on visual detection of prey, complete darkness is a serious handicap. If you get different results, what explanations can the students think of?

Experiment 9

Question:

Will spiders build webs in the cold?

Hypothesis:

If we put one jar in the refrigerator, the spider will be too cold to build.

Methods:

Put two web-building spiders of the same type in two jars. Put one jar in the refrigerator and leave one at room temperature. Does the one in the refrigerator build as well and as soon as the one at room temperature?

Result:

Your result is a description of your spiders' activities.

Conclusion:

Depends on the temperature of the refrigerator and the type of spider. Some spiders will build normally in the refrigerator (which is also dark). Can they catch prey in the refrigerator?

Experiment 10

Question:

Do particular kinds of spiders prefer certain types of prey? Must the prey be moving?

Hypothesis:

Whatever the students predict.

Methods:

You can do this with either a web-building spider in a web or a non-web-builder. Offer one spider two different types of prey at the same time and record which prey is taken. To be sure there really is a preference, you'd have to offer the spider the same choice of prey more than once. Do different kinds of spiders share preferences?

Result:

Your result is a statement of your spiders' choices, or abstentions.

Conclusion:

Prey must be moving. Spiders generally prefer flies or other flying or actively moving insects. They will sometimes accept wormlike animals like beetle larvae (grubs) or fly larvae, but not as readily. What about caterpillars, earthworms, or small slugs? Prey usually must be smaller than the spider.

A spider's hunger can affect its preferences. A hungry jumping spider has an abdomen that's smaller in circumference than its cephalothorax and looks caved in. Watch its abdomen plump out after it's fed. A hungry jumping spider will accept prey that it ignores when well fed.

Chapter 14
Praying Mantises: The "Smartest" Insects

Introduction

Mantises may be the most compelling insects I've ever met. They seem much too intelligent to be "only" insects. My family has been totally charmed by the many we've kept as pets. In this chapter I describe the features that, for me, set them apart from other bugs. I'll also describe their predatory behavior and how children can involve themselves in feeding a mantis. The chapter also outlines several experiments to investigate prey recognition, habitat choice, climbing behavior, and more.

Figure 3 in the Postscript provides a range of questions and answers comparing behavior of the predators discussed in Chapters 12 to 15. A list of the questions represented in Figure 3 is provided in Figure 4, which can be photocopied and handed out for the students to answer.

Materials

1. A mantis. You can order egg cases from a biological supply company (see Appendix) or catch one outside (instructions later in the chapter).

2. A terrarium at least about 6 by 6 by 10 inches (15.2 x 15.2 x 25.4 cm).

3. A lid for the terrarium that is not airtight.

4. Live prey for the mantis. This is not as hard as you might think. I'll tell you how to find suitable critters in the Field Hunt section. (See also Chapter 2 for suggestions on how to attract and trap bugs for the mantis.)

Background Information

At first glance mantises look a bit like spiders because of their long legs. Their appearance put me off at first until I

Fig. 14.1 An adult Chinese mantis on my back deck railing, looking at me. Her rounded abdomen is typical of females full of eggs. Note the triangular face and round eyes that give her a sentient look, and the spines on the two front legs that help grip prey.

watched one long enough to realize that the resemblance to spiders stops with the legs. In many ways their behavior is more like that of a cat or even a person.

Our first captive mantis, whom we eventually named Mantie, intrigued us the minute we brought her inside by turning her head to look at each of us in turn. I'd never seen another insect turn its head—it's odd how far this went in making her look intelligent and sentient, even sort of human. The head turning made her seem interested in us, which I think was not entirely an illusion. And her face looked recognizably like a face, which is not the case for most insects. Like all mantises, she had a heart-shaped face with two round eyes and a little pointed mouth (see Figure 14.1). She held her front legs folded in front of her as if she were praying. We learned in time that she used her front legs more as arms than legs, for holding prey and climbing, not for walking. This too made her seem less insect-like and more human.

Preying

Mantises are predators (they eat only live prey), which is where the catlike behavior comes in. Mantises don't actively search for prey as some predators do. A cheetah, for example, can chase down its supper, but a mantis can't walk very fast. A mantis hunts by sitting very still and waiting. If a mantis sees a cricket or a fly out of reach, it will creep slowly toward it. Its body stays very level as a cat's does; only the legs move, to avoid startling the prey. When the mantis gets within striking distance, it will sometimes sway gently back and forth before it strikes. I don't know why—to improve its depth perception? I asked a friend why baseball players sometimes sway back and forth before some impending action, like batting. He said, "You can move faster if you're already moving when you start." Maybe that's why mantises sway.

The strike from start to finish lasts just a fraction of a second. If you blink you'll miss it. A mantis doesn't take a step when it strikes; it merely leans forward and extends its long front legs lightning fast, grabs, and returns to the starting position. The front legs are jointed and clamp the prey in a scissor hold. Spines along the front legs hold the prey in place. The mantis takes small bites, eating the prey alive, and eats almost the entire thing from "head to toe." I have to admit watching the mantis eat is a little grisly sometimes.

Mantises will readily take prey from your fingertips. I've never had one grab my finger by mistake, but that's because I'm careful. The spines might hurt! I usually offer crickets when I feed a mantis by hand so I can hold the tip of the cricket's long outstretched legs and keep the cricket's body away well away from my fingers. The long legs are like a handle of sorts, because the mantis will grab the cricket's body. My own children will feed a mantis by hand, but I think it takes a good bit of familiarity with a mantis for a child to be so inclined.

To feed one by hand in front of group of students requires that the students stand absolutely still and quiet; otherwise the mantis is too distracted by the movement to eat. It helps to give the mantis a few minutes to get accustomed to having a crowd around (even a very still crowd) before offering it the prey. Familiarity over several days helps too—a mantis becomes more accustomed to students crowded around and watching, after you've hand-fed it in front of observers several times. (Pet stores usually sell crickets. In early autumn, crickets can be found under logs or brush. See Chapter 9 for more about procuring crickets.)

Feeding a mantis by hand is probably what I like best about keeping mantises. It's just icky enough to give me a thrill.

Grooming

Something else about mantises that I really like is how they clean themselves. A mantis will clean its legs with its mouth after a meal. Then it uses the front legs to clean its head and face, as a cat does. It cleans its back legs too. A mantis really reminds me of a cat with the tip of that back leg in its mouth. I'm not sure how they do it with no tongue, but they do.

Reproduction

Mantises mate in late summer or early fall. The male stands on the female's back, the tip of his abdomen curved up to join hers. (Kindergartners recognized this as "riding piggyback.") I've seen mantises stay in this position for a couple of days. The female may eat the male or just chew off his head after the mating is over.

A female full of eggs has a rounded underside to her abdomen. (A male's abdomen is slender and pencil-like.) Because she's heavy, she can't fly and is hence very easy to catch. Probably eight out of ten adult mantises I or others catch turn out to be pregnant females, which is fortunate for those of us who want egg masses.

The female lays her eggs usually one to two weeks after mating. She'll lay them in captivity readily, usually two or more egg masses. The mantis begins egg laying by expelling a wad of foam from the tip of her abdomen onto a twig or the side of the terrarium. The eggs are suspended in the foam, which hardens into an egg case. They will hatch in the spring if fertile. (See Chapter 15 on baby mantises for ideas about what to do with the hatchlings.)

All mantises die after the eggs are laid, in autumn, no matter how well you care for them. They may die abruptly or weaken and stop eating gradually over a period of a month or so. If you lay a dead mantis on its back in a dry and airy spot, it will dry out without any odor and can be kept indefinitely for viewing.

Defensive Behavior

I'm curious about what specifically triggers a mantis to react the way it does when frightened. If you move your hand toward an adult suddenly, it will open its wings to make itself look bigger and more menacing. It draws its upper body and forelegs back, poised to strike—somewhat like a snake. The mantis draws back much farther when frightened than it does before making a food strike. And it may strike at your hand—although not trying to grab it but apparently simply to bump it. Young ones do the same thing when frightened, but without opening the wings because they have none.

In some adult mantises I've seen a completely different reaction to being frightened, a reaction I think evolved to conceal them rather than to threaten the intruder. I had one mantis in particular that did this every time I took the lid off of her terrarium. She froze with her body held vertically and her forelegs extended vertically above her. This posture made her look much more like a stem, and I suppose in general makes a mantis less visible to potential predators. Many of the baby mantises I have now in an insect sleeve cage (see Figure 2.6) adopt this posture when I spray water droplets through the mesh top of the cage.

This type of behavior—cryptic posturing—occurs other places in the animal kingdom. Bitterns are a type of marsh bird with vertical streaks on the breast. When startled, the bittern will freeze with its head thrown back, so that the vertical streaks on its throat and breast blend in with the marsh grass. It's effective enough that I've never been able to spot one, after years of searching. Any behavior or physical feature that helps an animal blend in with its background can be called cryptic. The mottled brown skin of a toad is cryptic coloring, as are the spots on a fawn, the stripes on a tiger. It's easy to think of many examples of cryptic coloring, less easy to think of examples of cryptic posturing. Why do some mantises strike when startled, while others freeze in a vertical position?

How to Get and Keep Mantises

Several species of mantises have broad distributions ranging throughout most parts of the United States. If you want one, you can try catching one outdoors.

Most of the adults I've had I have simply stumbled on unexpectedly around the outside of my suburban house—on the porch, an outdoor windowsill, or a screen. Keep your eyes open all the time for one and tell your students to—especially in August, September, and October. I usually find two or three per summer. I've never purposely gone out in search of an adult and found one. I have, however, been successful in catching many very small and recently hatched mantises with a net. The best net to use is a long-handled sweep net with an opening of about 15 inches (38.1 cm). The mesh part of the net is almost 40 inches (1 m) long and cone shaped. You can order a sweep net from a biological supply company

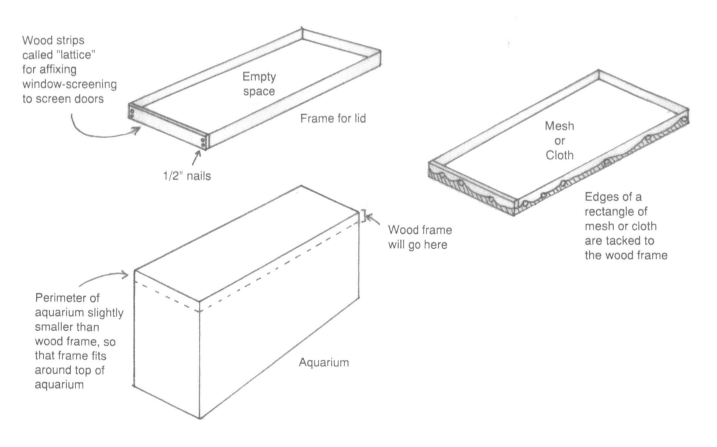

Fig. 14.2 Wood frame top for aquarium.

(see Appendix). Or you can use any net with a handle—a large fishnet or butterfly net. The best place to look is in areas of overgrown weeds that are waist high or taller. Sweep the net through the tops of the weeds quickly. The advantage of a sweep net is that the animals are swept to the tip of the long net and can't get out easily, giving you time to get your jar ready. The baby mantises jump—be fast! The babies (properly called nymphs) are also on twigs of shrubby trees in overgrown areas, on weeds along a creek, or along the edges of fields. They're climbers, so they're not likely to be on the ground. (An insect nymph is an immature insect of a species that undergoes "gradual metamorphosis" or "incomplete metamorphosis." For such insects, there is no larval and pupal stage. Rather, the hatching looks very much like a miniature adult, except that it lacks wings and sexual maturity. As it grows, it sheds its skin several times. After the last molt, it's a sexually mature adult with wings.)

If you can't find a mantis outside, you can raise your own from an egg case. (See Chapter 15 on baby mantises for a description of how to look for egg cases outside or how to order one.)

Keeping a single mantis is relatively simple. The terrarium I use is about 16 by 8 by 8 inches (40.6 x 20.3 x 20.3 cm). It should have a multibranched twig in it that doesn't rock for the mantis to climb on. A mantis who has nothing to climb is not happy. You'll need a lid to keep the mantis inside, but not an airtight one. A cloth lid is ideal. You can affix it to a wood frame or a rectangle that fits around the top of the terrarium (see Figure 14.2), or you can use a rubber band or just tape it on. The mantis won't try to squeeze under the edges so it needn't be tight. Keep the floor bare so the mantis can spot its prey. Any leaves or dirt on the cage floor will allow the prey to hide.

For a baby mantis the first enclosures should be small. A tiny mantis in a large terrarium will never be able to find its prey. For a mantis 1 inch (2.5 cm) long, an 8-ounce (.25 l) drinking glass is big enough. I don't use twigs for very small enclosures because a small mantis can climb the side of the enclosure. As the mantis molts and grows, get bigger enclosures.

A mantis can eat prey almost as large as itself. The easiest way to feed a large mantis is to set out a funnel trap (see Figure 2.3) baited with a quarter-sized piece of raw chicken, liver, or fish. You'll catch big flies. Chill them for two and a half minutes in the freezer, then dump them in the mantis's cage. They also like moths, crickets, beetles, grasshoppers—just about any insect. They don't like slugs, caterpillars, or roly-polies. I find field crickets by looking under boards or under sheets of plastic on the ground and especially inside piles of old damp grass clippings.

Camel or cave crickets hang out on the inside walls of sheds or basements. You can catch grasshoppers the same way I catch baby mantises, by sweeping a long-handled net through the tops of tall grasses. You can get moths by turning on your porch light at night. Use a plastic cup or jar to clap over one or whack it lightly with a rolled-up newspaper. A mantis will eat a wounded creature, as long as it's moving. I have had them eat freshly dead crickets that I've wiggled by the legs. They may even eat raw hamburger if you wave it in front of them with tweezers. Baby mantises like fruit flies. (See Chapter 16 for information on how to trap or raise fruit flies.) They also like *Tenebrio* beetles (adult mealworms), which are easy to raise (see Chapter 18).

A mantis will eat prey much smaller than itself, but if the prey is very small, make sure the mantis gets enough to eat. It will eat several times a day if given the opportunity, but twice a week is enough to sustain one.

Field Hunt

I don't recommend taking students on a field hunt for adult mantises, because you're unlikely to find one. They blend in quite successfully with natural surroundings and sit very still. Even if you look right at one in nature, you probably won't see it. I've spotted an adult mantis in its natural habitat (tall grasses and shrubby areas) only once. They're much easier to spot on human structures—a window screen, wall, deck, car, and so forth.

Show your students pictures of mantises and ask them all to look in August, September, and October. It's easy to catch an adult female with a jar. If you put the open jar slowly in front of her or over her, she'll probably step into it.

Mantis nymphs (babies) are fairly easy to catch with a net. I've never caught one longer than about $1^1/_4$ inch (3.2 cm) with a net. (Hatchlings are about $3/_8$ inch [1 cm]; adults of some species are up to 4 inches [10.2 cm.]) (See the Field Hunt section of Chapter 15 for information on finding young nymphs and egg cases.)

A Mantis at School

Getting Ready

All you need to begin is a mantis, a terrarium, a lid, a twig for climbing, and food.

Fig. 14.3 Sadie enjoys the first part of an adult Chinese mantis's walk on her right arm. Mantises always head for higher ground, so it starts up her arm. Sadie tries to transfer the mantis to her left hand, but it steps over instead and continues up toward her the top of her head.

Observations and Activities

What seems to impress children the most about mantises is holding them (see Figure 14.3). The first class I took a mantis to was a first-grade class. As the students sat gathered around watching me unsuccessfully try to feed the mantis, almost every one of them clamored and begged to hold it. They could see she was sitting very calmly on my wrist, and there was no menace involved. But there were too many of them to hand her around so we couldn't do it. After they all sat down at their desks, the mantis decided to eat, and I walked around to each seat so each child could see. One little boy, Kenny, put his head down on his desk and declined to look. But as I went to put the mantis away, there was Kenny. "Can I hold it, please?" Everyone else was on to some other

lesson. "Sure, Kenny." He flapped his hand out calmly and trustingly. I told him to hold his hand with the palm down, and put the mantis on the back of his hand. He recoiled only slightly when he felt her tickling feet. This was a full-grown Chinese mantis, the largest, at about 4 inches (10.2 cm). He stared intently as the mantis began her trek up his arm, as they always do in their yearning to climb. When it passed his elbow, he asked me to take it. Then he beamed at me from ear to ear. He was very pleased with himself.

A mother asked me to tutor her ten-year-old girl who was doing a school report on a mantis she had found. The girl's mother wanted her to interview me on the life history of mantises. The daughter dutifully asked a few questions and wrote down the answers, but with no enthusiasm. What she really wanted me to do was show her how to hold it to impress her younger siblings, her mom, me, and herself. I showed her how to put her hand down slowly in front of the mantis and allow it to step on. It usually will because your arm is something to climb. If you move too suddenly you'll frighten the mantis, and it may strike at you. The daughter accomplished her goal, easily and proudly. She even let it walk up the back of her neck, something most people can't tolerate (me included). I heard from her mom that the mantis was a big hit at school, and she was asked to take it around on her arm to some other classes. She was proud of that!

Holding Mantises by Hand

If you can manage it, it's probably a good idea to allow students who ask to hold the mantis to do so one at a time and supervised. Be sure to have students hold it on the back of their hands. This is important because when palms are up, fingers tend to curl upward, and the mantis may mistake a moving fingertip for prey, may strike at it, and even grab it. Those spikes on the mantis's front legs can draw blood. So—always keep palms down.

The easiest way to get the mantis on the back of a student's hand is to put the mantis on a desk, and position the student's hand, palm down, in front of the mantis. Because mantises instinctively climb, the mantis will almost always step on the hand and start moving up the arm.

Be ready to intercept it before it reaches the student's neck—most people don't like it on the neck. You can remove it just by putting your hand on the student's arm or shoulder in front of the mantis. As it steps on your hand, lift it away and put it back in the cage. If it's on your hand and you can't stop its upward motion toward your neck, simply lift your hand above your head. It won't walk down, but will stay on your hand until you can put it on a curtain or a desk or back in its terrarium.

A general rule of thumb for transferring a mantis is to place the object you want it to step on higher than the surface it's already on. That climbing instinct is always in play. If it gets on your hair or neck, don't freak out. A mantis can't grab anything bigger than a fingertip.

I've let many captives out of the cage in classrooms and they generally don't fly unless frightened. But a mantis may wander away if you leave it out for hours.

If you let the students feed it and watch it eat for several days, they may come up with some questions you can address experimentally.

Experiments

Remember that the hypotheses I give are just examples. Your hypotheses will be the predictions made by the class or a particular student. Your result for each experiment will be a statement of how your animals reacted to your experimental setup. Your conclusion is a statement of whether your prediction was confirmed or not. For each experiment, adding replicates increases your confidence in the validity of your conclusion, but they may be omitted if tedious for young children.

What characteristics of potential food cause the mantis to recognize it as potential food? People recognize food by sight and smell. Does a mantis recognize the shape and color of a cricket? Does it recognize the shape and color of all the animals smaller than itself that it could eat? That's a lot of different animals to recognize for an insect, even if it is a smart-looking insect. Is it something about the smell that they recognize? Do all prey make some sort of noise? The answer is something that students probably will not guess, although it is very simple, because it is so different from what we look for in food. There's one thing that all mantis meals have in common: They're all moving. Movement is what attracts mantises' attention, what signals their brains that the object before them is a potential meal. How efficient! It's much simpler to recognize movement than to recognize hundreds of kinds of creatures.

You can help students discover this on their own simply by either suggesting or seeing to it that the students offer both dead and live prey. If you tell them to bring in an insect and don't specify that it be alive, they'll bring dead ones. They may notice spontaneously that dead prey are ignored. If not, ask them if the prey that were struck at were dead or alive.

It's good to lead them at this point to a suspicion that being dead or alive has something to do with it. Conduct an experiment to get a definitive answer. Experiments 1 and 2 address the issue of prey recognition.

Experiment 1

Question:
    Will a mantis eat a dead insect?

Hypothesis:
    We think our mantis will eat any bug, dead or alive.

Methods:

Put a live bug and a dead specimen of the same type in the terrarium. Crickets are a good choice, because live ones move a lot. A live moth may sit still for hours. Watch until one is eaten or at least struck at by the mantis.

Result:

Your result is a statement of your mantis's actions.

Conclusion:

The prediction is not confirmed. A mantis will eat or at least strike at live prey, but it will ignore dead prey. If your mantis ignores the live prey too, it may be upset by too much activity around it, it may not be hungry, or it may object to that particular prey type for some reason. Keep trying.

Experiment 2

Question:

Do mantises ignore dead prey because they are dead and rotting, or because they are not moving?

Hypothesis:

We think our mantis ignores dead things because they smell bad.

Methods:

Hold a freshly dead cricket by the back legs with a pair of tweezers and jiggle it in front of the mantis. You may want to withhold food from the mantis for a day or two before you do this. You can kill the cricket painlessly by freezing it. This will leave it looking intact.

Freezing an insect is not as cruel as it sounds because they are cold-blooded (poikilothermic). They don't shiver and feel miserable; they simply slow down gradually until they stop.

Result:

Your result is a statement of your mantis's response to the cricket.

Conclusion:

The prediction is not confirmed. The mantis will eventually strike at the cricket and eat it. Mantises ignore dead prey because they are not moving. It may take persistence and trying different amounts of jiggling. A mantis that's overstimulated by little hands waving around it will usually not eat.

Try the same thing with a piece of hamburger, then a piece of apple. Does taste have something to do with acceptance of food after capturing it?

Why do mantises choose the places they do to wait for prey? Why don't we see them on the forest floor?

Experiment 3

Question:

Why do mantises climb? Will a mantis in a high place catch flying prey faster than a mantis in a low place?

Hypothesis:

We think a mantis in a high place will catch flying prey faster than a mantis in a low place.

Methods:

Offer a mantis a moth in a cage with twigs whose branches reach almost to the top. At another time offer the mantis a moth in a cage with nothing to climb on. Do it simultaneously if you have two mantises and two cages. Record how long it takes the mantis to catch the moth in each cage. Which mantis is the fastest? Repeat each trial several times if you want to be sure the results are not due to chance alone.

Result:

Your result is a statement of the time required to capture the moth in each situation.

Conclusion:

The prediction is confirmed. A mantis will usually catch a moth faster in a cage with twigs to climb on.

Experiment 4

Question:

Does climbing help mantises catch nonflying prey, like crickets? Does a mantis in a high place catch nonflying prey faster than a mantis in a low place?

Hypothesis:

We think climbing helps mantises catch all prey.

Methods:

Set up one terrarium with climbable twigs as in Experiment 3. Set up the other with no twigs. To be realistic you should put leaf litter on the floor of each cage, because a cricket in nature will hide. Offer a mantis a cricket in each cage and record how long it takes to capture the cricket in each cage. Then try the experiment again with a bare floor in each cage.

Result:

Your result is a statement of the time required to catch the cricket in each situation.

Conclusion:

The mantis will probably never catch a cricket in a cage with leaves on the floor. In the cage with a bare floor, it'll catch the cricket, but twigs won't make any difference. Climbing does not help catch nonflying prey.

Experiments 3 and 4 bring up the issue of different predator strategies. Mantises are sit-and-wait or ambush predators. This strategy works well for flying prey that might come to rest next to a mantis. It does not work well for concealed prey, like crickets. Many predators, such as towhees and other birds, will actively search through the leaf litter for prey, flipping over leaves with their feet. A mantis will never turn over objects to expose prey as a towhee will, and most insects on the ground are hiders.

Can the students think of other examples of either strategy (sit-and-wait versus actively search)? What strategy does an ant lion (Chapter 12) use? How about spiders? Frogs? Anteaters? Maybe the students can come up with other experiments to find how the mantis's strategy does or doesn't work with various types of prey. Can they think of a nonflying prey that doesn't hide?

Experiment 5

Question:

Does the twiggy shape and green or brown coloring of a mantis help it hide from its enemies?

Hypothesis:

We think the mantis's shape and color help it hide from its enemies.

Methods:

Release your mantis in a bush while the students aren't looking. At the same time, place a small red ball no bigger than an egg in the bush. Substitute anything that's a different shape and color than the mantis but not bigger. Ask the students, one at a time, to find the mantis and the ball. How many students spot the ball first? How many spot the mantis first? Does the mantis's shape and color help conceal it?

Result:

Your result is a statement of the number of students spotting the ball first and the number of students spotting the mantis first.

Conclusion:

The prediction is confirmed. A mantis in a bush is very hard to spot, for anyone. Those that stray onto a sidewalk or porch or the wall of a building are much easier to see.

Experiment 6

Can a mantis learn from experience? Specifically, will a mantis be able to catch crickets faster after a two-week diet of crickets? The ability to improve its efficiency as a predator would be to the mantis's advantage. But is its brain capable of learning? Ecologists recognize several components of foraging or prey capture: (1) search, (2) pursuit, (3) capture, and (4) handling (which includes eating). Many animals have been shown to be able to improve their efficiency, at each step, with experience. Obviously, a predator such as a lion, which learns predator skills in part from its mother, would improve with experience.

But what about an animal whose behavior is not learned from a parent but is completely instinctive? You can do a separate experiment with each component of prey capture. I've outlined an experiment to test the mantis's ability to spot prey faster with experience.

Question:

If I feed my mantis a cricket every day for a week or two, will the mantis spot the cricket faster at the end of the week than it did at the beginning?

Hypothesis:

We think the mantis will spot a cricket faster after a two-week diet of crickets.

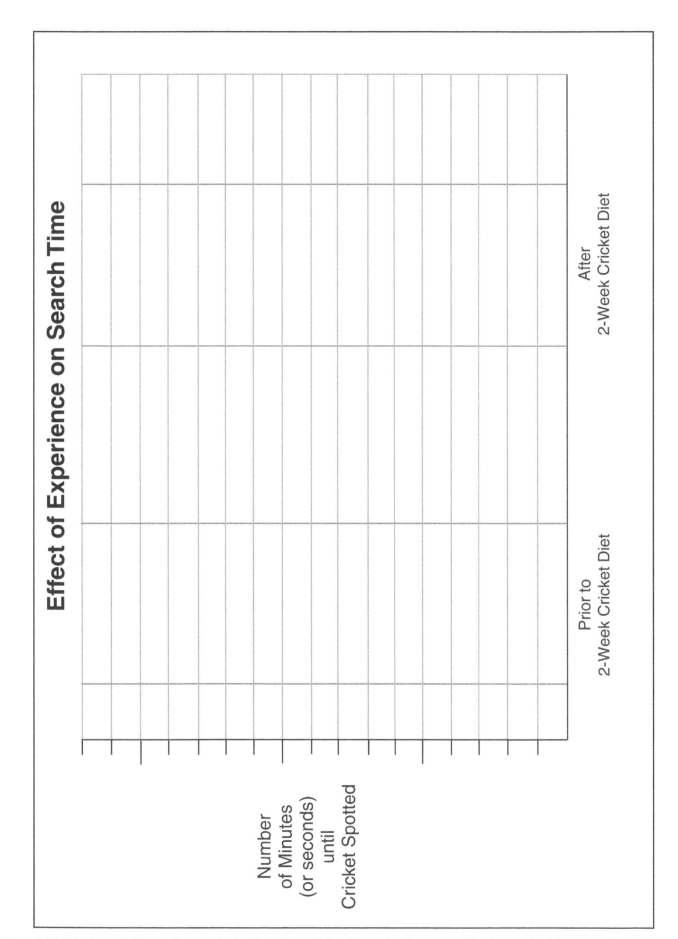

Figure 14.4 Blank graph for recording results for Experiment 6. (See Figure 7.4 for an example of a completed histogram.)

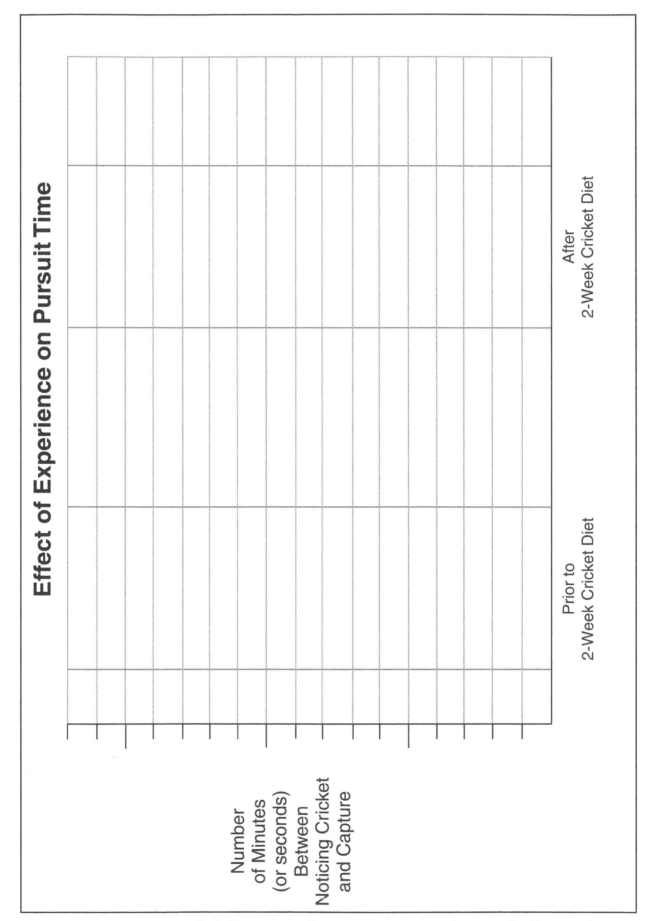

Figure 14.5 Blank graph for recording results for Experiment 6. (See Figure 7.4 for an example of a completed histogram.)

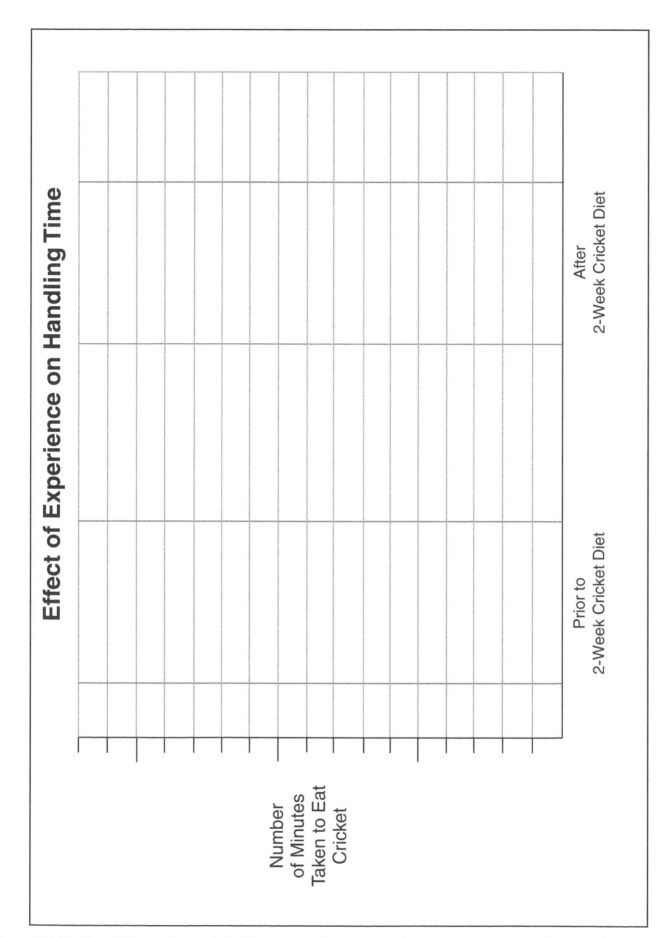

Figure 14.6 Blank graph for recording results for Experiment 6. (See Figure 7.4 for an example of a completed histogram.)

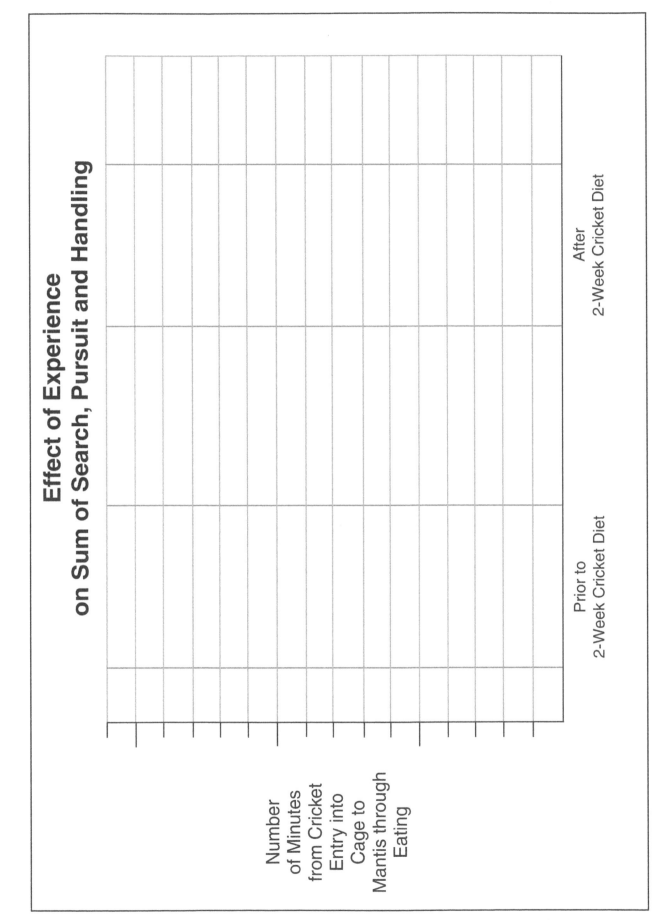

Figure 14.7 Blank graph for recording results for Experiment 6. (See Figure 7.4 for an example of a completed histogram.)

Methods:

At the beginning of a two-week period, and again at the end of a two-week period, record how long it takes your mantis to notice a cricket. Measure the time from when the cricket is first put in the cage to when the mantis notices the cricket. (Because the mantis can turn its head, it's obvious when the mantis is watching something.)

Use crickets that are more or less the same size for the tests and be consistent in how far from the mantis the cricket is introduced into the terrarium. Feed the mantis nothing but crickets in between the two tests.

Alternatively, you could measure the other components of prey capture: the time from when the mantis notices the cricket to when it captures the cricket (pursuit); the time from capture until the time eating is finished (handling); or the sum of search, pursuit, and handling. The capture itself is almost instantaneous for a mantis. Do the other components of foraging improve with experience?

Result:

Your result is a statement of your mantis's response. You can record your results on the histograms in Figure 14.4. Figure 7.4 provides an example of a completed histogram.

Conclusion:

I have not yet been able to confirm this prediction.

Experiment 7

It's easy to experiment with what type of prey mantises will accept. I've been surprised at some of the things they've refused to strike at, such as slugs and caterpillars. They'll grab roly-polies and bite them but then drop them. It would be interesting to see if they'll eat sow bugs, which look like roly-polies but can't roll up. I've seen a mantis outdoors eating a large wasp, which surprised me. Is the mantis's exoskelton impervious to stings? The largest animals I've seen mantises eat are other adult mantises and a $1^1/_2$ inch (3.8 cm) tough grasshopper. Here's a sample question.

Question:

Will a mantis eat a mealworm (a beetle larva; see Chapter 18)?

Hypothesis:

We think our mantis will eat a mealworm.

Methods:

Put a mealworm in the terrarium with the mantis.

Result:

Your result is a statement of your mantis's response. You can record your results on the table in Figure 14.8.

Conclusion:

The prediction is confirmed. A mantis will eat a mealworm.

Prey Offered	Prey Ignored	Prey Struck at and Dropped	Prey Eaten
Slugs			
Caterpillars			
Pill bugs			
Mealworms			

Figure 14.8 Blank table for recording results for Experiment 7.

Chapter 15
Baby Mantises (Nymphs)

Introduction

Praying mantises are large predatory insects with a distinctive appearance. Most insects use all six legs for walking, but mantises use the first two for grabbing and holding prey. When at rest, these legs are held folded in front, resembling hands held in prayer. The Chinese mantis, one of the largest and most common species and the one I've used most for the activities and experiments in this chapter, is 4 to 5 inches (10.2–12.7 cm) long at adulthood in late summer (see Figure 15.1).

Chapter 14 deals with adult mantises. I've put baby mantises in a separate chapter because I always cover them separately with classes. This is because adults are available only in the late summer and fall, and babies are generally available only in spring. I usually have only a couple of adult mantises at a time because they need large containers. But it's not difficult to keep thirty babies at one time. Each student can keep one in a small container at his or her desk. The activities and experiments involving baby mantises focus on their growth and molting, and their effect on one another. Most of the experiments involving adults are related to their predatory behavior. The babies are predators too, of course, so many or all of the experiments described for adults can also be done with babies. But most of what I describe for babies can't be done with adults.

Baby mantises are generally called nymphs. (See Figure 15.2.) Mantises mature through "gradual metamorphosis," which means there is no larval or pupal stage. Insect species that develop through gradual metamorphosis have young that look very similar to the adult, except that they lack wings and are not sexually mature. As the immature stages grow, they molt or shed their exoskeletons several times. The word *nymph* refers to the young at any stage of this development, from hatching to sub-adulthood. I refer to the young in this chapter as "babies" because I'm talking about very young and very small nymphs, and because children are more intrigued by them when I refer to them as babies. They are babies, if babies can refer to any animals that have just hatched or been born.

Figure 3 in the Postscript provides a range of questions and answers comparing behavior of the predators discussed in Chapters 12 to 15. Figure 4 provides a list of the questions represented in Figure 3, which can be photocopied and handed out for the students to answer.

Materials

1. Baby mantises. To have enough for a whole class, you'll probably need an egg case, which you can order from a biological supply company (see Appendix). You may be able to order one from a garden supply company (as predators of garden pests). You can possibly find one outside. You may be able to catch a lot of baby mantises with a sweep net outside, but probably not thirty. (See the Field Hunt section in this chapter for more about finding eggs and catching babies.)

2. One *Drosophila* culture vial for each baby mantis. The vials are clear plastic cylinders 4 by $1^1/_4$ inches (10.2 x 3.2 cm), which can be ordered from a biological supply company. (*Drosophila* are fruit flies and are used commonly in biology labs, so even a science hobby shop may have vials and plugs.) Small clear jars such as baby food jars will do, although vials are easier. Some spices and the tiny colored candy for sprinkling on cakes come in clear plastic vials that are very similar to *Drosophila* vials. You could ask each parent to donate one.

3. Foam rubber *Drosophila* vial plugs, which can be ordered from a biological supply company.

4. One cotton swab for every vial to provide water.

5. Fruit flies or aphids or other small insects to feed the mantises. Fruit flies can be ordered, cultured, or trapped (see Chapter 16 for information on flies, Chapter 17 for aphids). Wingless fruit flies from a biological supply company are the easiest prey for very small mantises. Mantises more than two to three weeks old can easily eat wild fruit flies you can trap.

Background Information

Mantises are fascinating creatures. They seem very intelligent and display an array of behaviors that seem very un-insectlike. They groom themselves like cats. They turn their heads and watch us like people. They use their "arms" to hold their food like monkeys. They stalk and capture prey like the most ferocious predators. The larger ones are easier to observe, but the babies do all these things too. Adults move slowly and deliberately and are easy to hold; babies run fast and jump, so they are harder to handle without losing them. They are easier to handle as they grow.

The easiest way to get baby mantises is to get them in an egg case before they hatch. Adult mantises lay their egg cases in the fall, before they die. The eggs normally hatch in late spring, although egg development can be accelerated by keeping them indoors. Each egg case will yield fifty to one hundred babies or nymphs. There is no way (that I know of) to distinguish their gender before adulthood.

Fig. 15.1 Adult Chinese mantis with arms outspread, on my deck railing.

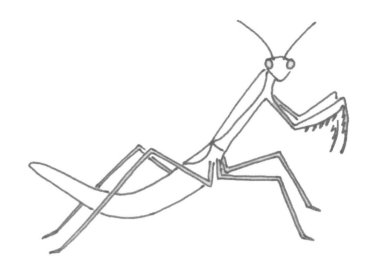

Fig. 15.2 Drawing of a baby mantis or nymph. The nymph, which lacks mature functional wings, is ¼ to ³/₈ inch (0.6 cm–1 cm) in length at hatching, up to 3 inches (7.6 cm) or more before the final molt to adulthood.

Watching an egg case hatch is a thrilling experience. The babies come squeezing out head first through cracks in the case. Less than ³/₈ inch (1 cm) in length, they look like white worms or skinny fish with two black eyes. The headfirst entry into the world and the wet and squeezed appearance of the new arrival remind me somewhat of a human birth. But they keep coming and coming! Dozens ooze out and dangle upside down en masse as their newly exposed exoskeletons harden and their skinny little white legs unfold. Within minutes, they're off and running, one by one, until the terrarium is filled with miniature mantises.

The baby mantises won't eat for a day or two after hatching; then they will eat fruit flies, aphids, or other small insects. They strike at and grab their prey with their spiny forearms lightning fast, just as the adults do. The prey is held tightly with one or both forearms and munched slowly until nothing remains.

Molting is probably the most remarkable and conspicuous feature of captive hatchlings. An insect's skin, more properly called the exoskeleton, is stiff and provides support for its body. The exoskeleton doesn't grow, so the animal must shed the exoskeleton as its body grows. A new, soft one forms underneath before the old one is shed, or molted. After the new exoskeleton is exposed, it expands a little and then hardens. Mantises molt about six to nine times before reaching adulthood. In my experience, their first molt (after the molt that occurs during hatching) comes at the age of about twelve days if they are fed fruit flies more or less continuously. The growth rates of cold-blooded (poikilothermic) animals vary a great deal depending on quantities of food and temperature. The time to the first molt can vary considerably too, since it depends on the growth rate. (Cold-blooded is actually a misnomer because they may not be cold at all. The proper term, poikilothermic, means that they cannot regulate their body temperature internally as we homeotherms do, but must regulate it behaviorally by sitting in the sun or shade, etc.)

How to Get and Keep a Baby Mantis

I order egg cases from a biological supply company, or use an egg case from a female captive of the year before. The egg cases are usually attached to a twig or the side of the terrarium and should be kept suspended so that the hatchlings can hang upside down from it as their new exoskeletons harden. You may be able to find one outside (see the Field Hunt section in this chapter).

The hatchlings will begin to disperse as soon as their exoskeletons harden, so the egg case must be kept enclosed in a terrarium or other enclosure with a cover that allows at least a little air to circulate.

Most mantis egg cases hatch over a period of one to two hours. Some, like the egg case of the Carolina mantis, produce one or two hatchlings every couple of days for a couple of weeks or longer. There are several species of mantises in the United States, and the number of hatchlings can vary from the usual fifty to one hundred. If you want your students to keep individual mantises alone in vials at their desks, you can remove the hatchlings from the terrarium as soon as they are moving around. A small mantis will step into a vial if you goose it from behind with a cotton swab or a vial plug.

They can be kept together safely for a while but will eventually begin to eat each other. How soon depends on how well they are fed and how crowded they are. I kept ten or so in a container the size of a hatbox (an insect sleeve cage, see Chapter 2) for two months before I saw any cannibalism. Other times I've seen cannibalism within three to four days of hatching.

Feeding Baby Mantises

The easiest way to feed baby mantises is by raising your own fruit flies. It takes about six vials of breeding fruit flies to feed thirty vials of baby mantises for two weeks—maybe more. (See Chapter 16 for instructions.)

To feed baby mantises in a terrarium, just dump fruit flies into the terrarium every day. One fly per mantis per day is enough. If the fruit flies are winged, you'll need to chill the fly vial in the freezer first for sixty to ninety seconds to paralyze the flies temporarily so they won't fly away during transfer. If they are vestigial winged fruit flies, no chilling is necessary. (This flightless and almost wingless variety has to be ordered.)

Most of the baby mantises will hang upside down from the lid of the terrarium, which can be a problem at feeding time. If your terrarium is crowded (fifty to one hundred babies), some of them will escape every time you lift the lid. Thinning the population can help. Escapes can be avoided altogether by keeping the mantises in an insect sleeve cage instead of a terrarium. These can be ordered from a biological supply company (see Appendix) or made with a household bucket and piece of cloth (see Chapter 2 for an illustration and instructions).

If your mantises are in separate vials, you can feed them by tapping two or three flies into each vial. You'll have to chill the flies first if they're winged, as described earlier. If they're vestigials or flightless flies, no chilling is necessary, but you need to smack the bottom of the fly vial before opening it to knock all the flies to the bottom. Then you can quickly tap out a couple of flies at a time.

You can trap fruit flies easily when it's warm, but, unless you have several traps, probably not enough to feed thirty baby mantises. All you need to trap flies is a jar and some smelly fruit (see Chapter 16). Half a grapefruit works well after someone has eaten the fruit, leaving just the membranes and the rind. Bigger fruit flies that you may catch in traps need to be chilled for 150 to 180 seconds.

Baby mantises like aphids, so if you can find a weed outside with aphids you may have an endless supply of food (see Chapter 17). Aphids don't move much, though, so it sometimes takes the mantises a while to notice them.

After the mantises have molted once or twice, they're big enough to handle very small cricket hatchlings. In the spring, I find baby crickets easily under boards, under garbage cans, and so forth. A piece of cardboard will nudge them into a jar. Most get away, but if I come back in an hour they're back in position.

Watering Baby Mantises

To provide moisture to mantises in a terrarium, spray the inner walls once a day with small water droplets from a plant sprayer. Hatchlings will get stuck and drown in big drops. To water those in vials, keep the wet end of a cotton swab in the vial, the central shaft of it held in place between the stopper and the vial, with the dry end sticking out of the vial (see Figure 15.3). It's easy to remove the cotton swab every day to rewet it, without removing the stopper. The students can do this easily (see Figure 15.4). When you have flies in there you won't want to remove the stopper. Be sure the mantis is not standing on the swab before you remove it.

If you're using baby food jars instead of vials, use cloth lids secured with rubber bands. You can water the mantises by moistening the cloths.

Fig. 15.4 *Tyewanda and Austin dip their cotton swabs in water before replacing them into their vials.*

spot them in the grass. (A sweep net is a long-handled net for catching insects [see Figure 15.5].) The mesh part of the net is almost a meter long and has an opening of about 15 inches (38.1 cm). The length enables you to pinch it shut and keep insects trapped while you get your

Fig. 15.3 *Drawing of a* Drosophila *culture vial with a cotton swab in place.*

Field Hunt

At the right spot and the right time, baby mantises are easy to catch with a net. I live in a suburban area, but behind my house is a ditch and an area of overgrown grasses and wild shrubs. If I run a sweep net quickly through the top of the waist-high grass in late spring, I get a lot of baby mantises in the net even though I can't

Fig. 15.5 *Alan and Sadie with a sweep net.*

jar ready to receive them. Look for bees or wasps inside the net before you grab it!

Egg masses are found in the same sorts of places. They're stuck to the top or middle of tall grasses and stems, usually at knee or waist height. Egg masses start out as foam, deposited around a stem. The eggs are embedded in the foam, which hardens to protect them. The result is a light brown, walnut-sized, irregularly shaped mass with a rough irregular surface (see Figure 15.6). If you find a perfectly smooth and round thing about the same size, it's an insect gall and not a mantis egg case.

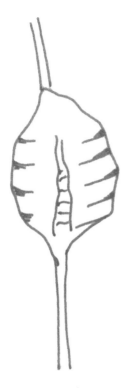

Fig. 15.6 A mantis egg case, about 1 inch (2.5 cm) tall.

Carolina mantis egg cases are smaller and conform somewhat to the shape of the stem. They are very hard to spot in nature.

Baby Mantises at School

Getting Ready

Start your six vials of fruit flies two weeks before you plan to put the mantises in individual vials, because it takes about that long to go from egg to adult fly (see Chapter 16 for instructions). If you have a bunch of

mantis hatchlings and no flies, you can order the flies from a biological supply company and probably get them in just a few days. Most of the mantises will survive a few days without food.

If you plan to let each student have his or her own mantis, you need a vial with a stopper or some alternative enclosure and a cotton swab for each one. The mantises should be in the vials when you give them to the students, unless you're willing to have a lot of mantises escape. Caution the students about twirling their cotton swabs. I had a student once who twirled his swab absentmindedly while his mantis was standing on it. The mantis got wrapped around the swab and flattened and poor Christopher was horrified.

You can feed the mantises yourself, but children age eight and up can learn to do it easily enough. I find it works better to have the students transfer a couple of flies to an empty vial first, and from there into the mantis's vial (see Figure 15.7). That way they don't have to worry about containing the mantis and a vial full of flies at the same time. Anytime you or they get ready to open a vial, whack the bottom of it hard first to knock the mantis or flies away from the top. Otherwise mantises and flies will run out (see Figure 15.8).

Observations and Activities

If every child has his or her own mantis, just watching them eat and molt is interesting enough that you can stop there if you want to. A first-grade class and a third-grade class I gave mantises to last year kept them for two weeks, which worked out well. The vials stayed in the grooves in the desks intended for pencils. After two weeks the students showed no signs of losing interest, but the feeding routine can get tiresome if the teacher is doing it all. You may also run out of flies after two weeks, unless you started six more vials two weeks after you started the first six. I handed the mantises out when they were about a week to ten days old. During the two weeks that the students had them, all but one of the mantises molted, although fewer than half of the students actually saw the mantis in the process of molting.

The most exciting aspect of keeping baby mantises captive is watching them molt. Explain to the students that we, and all vertebrates (fish, mammals, birds, amphibians, and reptiles), have skeletons on the inside. Some of the students may be able to tell you why we have skeletons—to support our bodies. Explain that an insect has a skeleton outside its body. Its skeleton is a stiff skin that supports its body. *Exo* means "outside."

Fig. 15.7 Tyewanda and Austin prepare to tap a couple of fruit flies from a culture vial full of flies into an empty vial.

They may guess a couple of other animal groups that have exoskeletons. Several students in the first-grade class I worked with suggested snakes as an example, and that's good thinking, but snakes shed their skins, not exoskeletons. Snakes have an internal skeleton like we do; a snake's skin provides no support.

The first time a child (or anyone) sees an insect in the process of molting, he or she is likely to be very impressed. My most vivid memory of sharing baby mantises with children is of the morning I discovered first grader Dulce watching her mantis molt. Hers was the first in the class to molt. She was grinning from ear to ear and ran to greet me as I entered the room. I asked her to tell me exactly what she saw. She said the head came out first and it was all white. At first she thought it was dying, but then she realized what was happening.

Watching the mantises eat is fun too. If you keep the mantises in the class for two weeks, almost all of the students will see the mantis in the process of eating a fruit fly (which takes a minute or two), but very few will see it strike or grab the fly (which takes a fraction of a second). The mantises often eat first thing in the morning because the children's arrival disturbs the flies. The flies' movement causes the mantises to notice them. Many students will see the mantis eat every day yet will continue to find it interesting every time. When children are interested in something, their powers of observation are amazing. There was a little boy in this first-grade class named Sam who attracted my attention because he was young and seemed in some ways unready for the

academic and social challenges of first grade. He was uncomfortable in the class. One day as I put new flies in the vials Sam told me that he'd noticed his mantis had bumps on its arms. He was referring to the spines on the forelegs. I asked him what he thought those bumps might be for. He said, "To help it hold on to its fly so it can't get away." I had not planned to even mention the spines to the students because I thought they were too tiny to see. Sam was one of those who surprised me. He was the only one in the class who noticed the bumps, let alone figured out their function. It was fun to hear him tell the class his discovery!

It might be a good idea to have a photograph of an adult mantis with something to give it scale, as well as an egg case for the students to look at. Without something to remind them, the students forget that their own captives are babies. Aphids molt too and if you're using aphids as mantis food, the aphids will leave tiny white exoskeletons in the vial. The students will think these little white specks and any dead aphids are their mantises' eggs, unless they have some concrete reminder that their mantises are babies themselves.

You can use the table in Figure 15.9 to record daily tallies of the number of mantises seen eating or molting.

Fig. 15.8 Mena and Clint round up a couple of escaped mantises.

Experiments

Remember that the hypotheses I give are just examples. Your hypotheses will be the predictions made by the class or a particular student. Your result for each experiment will be a statement of how your animals reacted to your experimental setup. Your conclusion is a statement of whether your prediction was confirmed or not. For each experiment, adding replicates increases your confidence in the validity of your conclusion, but they may be omitted if tedious for young children.

Experiment 1

Question:

Will a mantis that is fed more molt sooner?

Hypothesis:

A mantis that is fed more will molt sooner than one that is fed less.

Methods:

Have two or more mantises from the same egg case. It's important that they be the same age. Feed one mantis six flies or more per day every day. Feed the other mantis one fly every other day, which is certainly enough to keep it alive. Record which molts first. Use more than one pair to have confidence in your results.

Result:

Your result is a statement of the date each mantis molted.

Conclusion:

The prediction is confirmed. Mantises that eat more should molt sooner, because they grow faster. A mantis or any insect molts when it outgrows its exoskeleton.

A variation of this is to feed one mantis more consistently until adulthood. The mantis that is fed more should mature at a larger size and molt more times (about nine times) than one that is fed less (about six molts).

Experiment 2

Question:

Will a mantis that is kept warmer molt sooner?

Hypothesis:

A mantis that is kept warmer will molt sooner. (The students may have no idea what to hypothesize here. An unrealistic hypothesis is fine.)

Methods:

You need two very young mantises the same age and size. Keep one in a cold place. Anything down to 40 degrees F (4.5° C) is all right. Keep the other one in a warmer place, anything up to 90 degrees F (32° C). A temperature difference of only 10 or 20 degrees F (–12 to –6.5°C) may be enough to make a difference. Feed them both adequately and equally (at least two flies per day). Which one molts first?

Result:

Your result is a statement of which mantis molted first. You can record your results on the table in Figure 15.10.

Conclusion:

The prediction is confirmed. The one kept warmer should molt sooner, because warmth speeds the development of mantises and all other "cold-blooded" or poikilothermic animals.

If both mantises are kept hungry, the warmer one may not grow faster. A lack of food may limit its growth rate.

Experiment 3

Mantises are normally solitary creatures. They disperse soon after hatching and probably never see each other again. If forced to stay together in one enclosure, the hatchlings will eventually eat each other. The only time they come together as adults is to mate, and the females may eat their mates if they are hungry. So we might expect that two mantises cooped up together would be wary of each other. Are they? How can we tell if they are aware of one another or distracted by one another? One way is to see if they eat normally when another mantis is around.

Question:

Do mantises interfere with one another's prey capture?

Day (starting the day you get them)	Number of Mantises Seen Eating	Number of Mantises Seen Striking	Number of Mantises Molting	Number of Mantises Seen in the Act of Molting
1				
2				
3				
4				
5				
6				
7				
8				
9				
10				
11				
12				
13				
14				
15				
16				
17				
18				
19				
20				
21				
22				
23				

Figure 15.9 Blank table for recording daily tallies of students' observations.

The Effect of Cold on the Growth Rate of Mantises

Date Molted
(average if a group)

Mantis 1 or Group 1
(kept cold)

_____

Mantis 2 or Group 2
(kept warm)

_____

Which one or which group
molted sooner?

_____

The Effect of Interference on the Feeding Rate of Mantises

	Vial Containing One Mantis	Vial Containing Two Mantises
Number of flies consumed over _____ hours	a. _____	b. _____
Number of flies you would expect to be consumed in the vial with two mantises if they each ate as much as the solitary mantis		c. _____ (c is equal to 2 x a)

Did the two mantises housed together interfere with one another?
If b < c (significantly less than), then the answer is yes.
If b = c or b > c, then the answer is no.

Figure 15.10 Blank table for recording results for Experiments 2 and 3.

Hypothesis:

A mantis will eat fewer flies if another mantis is present, but not because the other mantis eats the flies.

Methods:

Before starting the experiment, you need to determine how long it takes a single small mantis (of the size you have) in a vial to eat ten fruit flies. Let's say you determine that it takes about twenty-four hours. For the experiment, you'll need at least two vials. Into one vial put two small mantises and twenty fruit flies. Close the vial with a sponge (foam rubber) stopper. Into the other vial put one small mantis and ten flies and close the vial with a foam rubber stopper. Push the stopper down into the second vial so that the single mantis has only half the space available in the first vial. You want to be sure that the amount of space available per mantis is constant. Then you'll know that the amount of space available is not responsible for the results. Leave the mantises in the vials for twenty-four hours, or whatever time you determined. Then count how many fruit flies are left in each vial.

To be confident that your results are not due to chance alone, you may want to set up several replicates of each situation.

Result:

Your result is a statement of the number of flies left in each vial after twenty-four hours. You can record you results on the table in Figure 15.10.

Conclusion:

The prediction is confirmed. When I've done this experiment I've found that the mantis alone consumed more flies than either of the two placed together. For example, after the predetermined number of hours the single mantis had consumed eight flies, while the two mantises together had consumed only a total of three flies between them. If they were eating at the same rate as the single mantis, they should have eaten sixteen flies together. The proximity of another mantis does seem to interfere with normal feeding behavior.

I'm guessing that, in this experiment, the motion of the other mantis in the jar distracts them so that they spend less time watching and pursuing prey. Mantises are, by nature, attentive to motion.

Insect
Reproduction

Chapter 16
The Two-Week Life Cycle of Fruit Flies

Introduction

Perhaps the most interesting feature of fruit flies is that you can see the entire life cycle, from adult parent to adult offspring, in two weeks. I don't know of any other insect that reproduces that quickly in captivity. What's more, they'll go through the whole cycle twice (to adult grandchildren) in a vial that fits in the palm of your hand, with no maintenance or notice required whatsoever. There's something compelling about this little self-contained world. Fruit flies are handy prey for many of the predators in this book.

The experiments herein address various factors affecting development, food requirements, and more.

Figure 6 in the Postscript provides a range of questions and answers comparing the life cycles of the organisms discussed in Chapters 16 to 18. A list of the questions represented in Figure 6 is provided in Figure 7, which can be photocopied and handed out for the students to answer.

Materials

1. One-quart or 1-liter jar for making a fly trap. (Three jars for Experiment 1.)

2. Rotting fruit for the fly trap.

3. Fruit fly vials and stoppers from a biological supply company (see Appendix) or other small containers like baby food jars. One per student or per group, plus at least five for your stock populations.

4. Cloth lids for baby food jars and rubber bands to secure the lids.

5. Dried banana flakes from a biological supply company (see Appendix) or ripe bananas.

6. One packet of active dry yeast, the type used for making bread.

7. Freezer for Experiment 1.

8. Alternative foods for Experiment 3.

9. Refrigerator for Experiments 4 and 5.

10. Dark room for Experiment 7.

11. Terrarium and lid and lamp for Experiment 8.

12. Dark cabinet for Experiment 9.

Background Information

Most of us think of a housefly when we hear the word *fly*. But did you know that mosquitoes are flies too? Flies are actually a huge and diverse group. Those I work with most are fruit flies, because they're easy to trap and easy to maintain in captivity. They reproduce happily in small containers, and their food doesn't smell bad. Many of the various flies that resemble houseflies lay their eggs on rotting meat so you'd have to provide that to raise them. They sometimes carry dangerous bacteria on their feet from landing on carrion and feces. My office mate (and future husband) in graduate school was studying the community of flies that breed on carrion. He had to have an area that was recessed into the wall with an exhaust fan to keep them indoors, and still the stench pervaded our lab. So, I stick to fruit flies. They are attracted only to fruit. Although rotten fruit has bacteria in it, it's not dangerous bacteria.

Flies are insects, of course, so they have three body parts (head, thorax, and abdomen) and six legs, as all insects do. Being insects, flies have the same type of life cycle as most insects. They go through what's called simple or complete metamorphosis. From the egg hatches a wormlike larva. A fly larva is often called a maggot. The

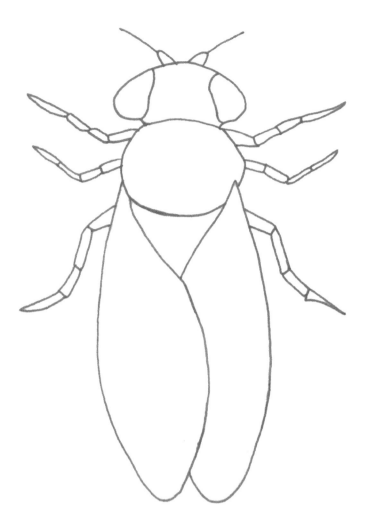

Fig. 16.1 Drawing of a fruit fly. Length $^1/_{16}$ to $^1/_8$ inch (1.6-3.2 mm), depending on species. They are yellowish or brownish in color.

larva eats and grows, shedding its skin, or exoskeleton, several times. After it reaches a certain size, it leaves its larval food source and pupates. Some insects, such as moths, spin a cocoon around themselves to pupate, but the fly's larval skin hardens into a tough casing instead. Inside that casing the body of the larva transforms into that of an adult fly. This stage of transformation is called a pupa. The pupa does not move around or eat. A fly pupa looks like a tiny fat cigar. When the adult fly inside is ready, it emerges from the pupal casing.

The wings are crumpled when the fly first emerges, but as blood is pumped into their blood vessels, the wings straighten out and harden. Then the adult can fly. The period following emergence is a very vulnerable period for winged insects, because they can't yet fly to escape predators. The life of an adult fly is usually brief, only a couple of weeks. Within a day or two of emerging, it seeks another adult fly of the opposite sex. The male courts the female to coax her into accepting him. Courtship behaviors vary according to species,

but usually involve wing movements, foot movements (tapping the ground or the female), and/or walking in circles or semicircles around the female. The species I have the most of (*Drosophila melanogaster*) are tiny ($^1/_{16}$ inch [1.6 mm]), but I can easily see the male following the female around vibrating his wings. If the female is not ready to mate, she may fly away from the male's pursuits or ignore him.

If she is ready, she will allow the male to stand on her back as they mate. When she is full of eggs, her abdomen will look swollen and white and easy to see, even though they are so tiny. She'll lay her eggs when she has rotting fruit to lay them on. (I'll describe in the next section how to provide a suitable place for egg laying.)

The fruit flies you get when you order from a biological supply company are *Drosophila melanogaster*. In addition to their tiny size, you can identify them by their bright red eyes. When biologists refer to fruit flies, they generally mean *Drosophila melanogaster*. If you leave a bowl of fruit sitting out on your kitchen counter and notice tiny flies on it that scatter when you pick up a piece of fruit, they are most likely *Drosophila melanogaster*. They are referred to also as the red-eyed fruit fly or the common fruit fly.

There are many other species in the family of fruit flies, Drosophilidae. Once you've seen a red-eyed fruit fly, you'll recognize the others by their similar, almost identical, body shape (see Figure 16.1), even though other species lack the red eyes and may be twice as large. When looking them up in a field guide, look up *Drosophila*, because field guides sometimes use unfamiliar common names, such as pomace flies.

Fruit flies are commonly used in biology labs because their generation time is so short (fruit doesn't last long in nature), and they have a lot of offspring. This makes them convenient for studies of genetic mutations—how subsequent generations are affected by genetic changes and so on. They also have an unusually large chromosome that's more easily studied with microscopes than the chromosomes of other organisms.

How to Get and Keep Flies

You can order fruit flies (*Drosophila melanogaster*) from a biological supply company (see Appendix), or you can catch your own. Catching flies is something I do on a daily basis in spring and summer, not because I love flies but because I feed them to so many other captives. Just about all the predatory insects in this book will eat fruit flies, with the exception of ladybug beetles.

The traps seem too easy to work as well as they do. I put a piece of rotting fruit in the bottom of a 1-quart or 1-liter plastic peanut butter jar. I lay the jar on its side with the lid off and the opening slightly lower than the bottom so rain won't collect in the jar. It works best when placed near a garbage can or compost pile, or near some other already existing rotting vegetation. When the weather is hot, the jar needs to be in the shade or the fruit dries out too quickly. Aging grapefruit rinds work best for me, although banana peels, apple cores, or canta-loupe rinds work too. Leave the jar for twelve to twen-ty-four hours. When you return, creep up on it slowly. Very slowly put the lid on the jar without touching the jar in any way, otherwise the flies will see you or will feel the jar move and exit before you can trap them. You may find from five to one hundred fruit flies of various sizes in the jar. If you don't find any, leave something large like a watermelon rind or several grapefruit rinds out in the open for several days, then remove them and leave the trap in their place.

When setting up the trap, you may want to put an inverted funnel of cardboard in the mouth of the jar to funnel the flies in toward the bait and make exit difficult. This isn't necessary if you move slowly when closing the jar to trap the visitors, but sometimes children like making the funnels. When the rotting fruit is dried out, it needs to be replaced.

Now you have a jar full of flies. What do you do with them?

If you just want to examine them, you can anesthe-tize them temporarily by putting the jar in the freezer for ninety seconds or so. If that's not long enough, check every twenty to thirty seconds after that until they stop moving. Alternatively, you can anesthetize them by putting some dry ice in the jar until they stop moving. Flies can be easily moved around with a fine paintbrush.

You may want to feed your flies and provide what they need to reproduce so you can observe their life cycle. That's really the interesting part. I use fruit fly culture vials from a biological supply company (see Appendix). They're plastic cylinders, about $1^{1}/_{4}$ inches (3.2 cm) in diameter and 4 inches (10.2 cm) tall, that come with a sponge plug. The sponge allows air to circulate while keeping most of the moisture and all of the flies inside. I use the vials for a lot of other bugs as well—praying mantis hatchlings, aphids, ladybugs, and small spiders. They're handy for keeping in a shirt pocket on a walk in case you spot an interesting bug. I get a lot of requests from students for the vials. Kids enjoy them because they look exotic and scientific and are fun to handle.

You can also use baby food jars or any small jar with a cloth cover secured with a rubber band. Among other things, fruit flies readily eat yeast that grows on bananas. To feed them in the culture vials, put in a

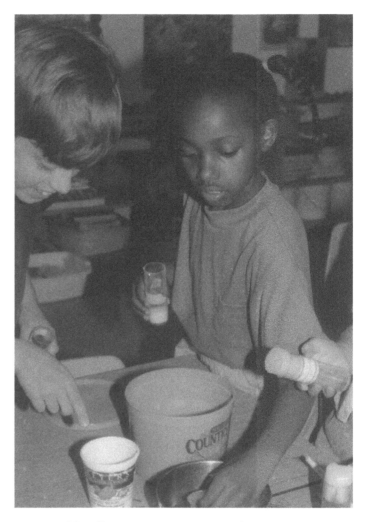

Fig. 16.2 Tanielle scoops up water to put on her banana flakes while Zach wonders about the yeast grains in the yogurt cup.

mixture of bananas and yeast before you put in the flies. I buy banana flakes from a biological supply company for this purpose (see Appendix). One tablespoon (15 ml) of flakes and 1 tablespoon (15 ml) of water make about 1 inch (2.5 cm) of goop in the bottom of the vial (no stirring necessary). Sprinkle about ten grains of yeast over the banana mixture. Don't add too much yeast or the mixture will rot and start to smell bad. You can make your own culture medium by mashing about 1 table-spoon (15 ml) of banana with about $^{1}/_{8}$ teaspoon (.63 ml) of water and putting it into the bottom of a vial. Add yeast as described. One disadvantage of using mashed banana is that it will get moldy eventually. The banana flakes have an antimolding agent mixed in. If you use baby food jars instead of vials, cloth won't keep moisture

Name _____

Tiny white worms called larvae will hatch from your flies' eggs. The larvae will eat your banana mixture. You may be able to see them. Soon they will crawl up onto the clear plastic. Then you can see them moving on the plastic. They will turn into pupae on the plastic. Pupae look like tiny brown cigars. Pupae do not move. An adult fly will come out of each little cigar, or pupa. All this will happen in the next two weeks.

Write down how many larvae and pupae you see every day for the next two weeks as best you can. If you can't count them all, just write "many." Larvae are white worms; pupae are brown cigars.

DAYS
after starting vial

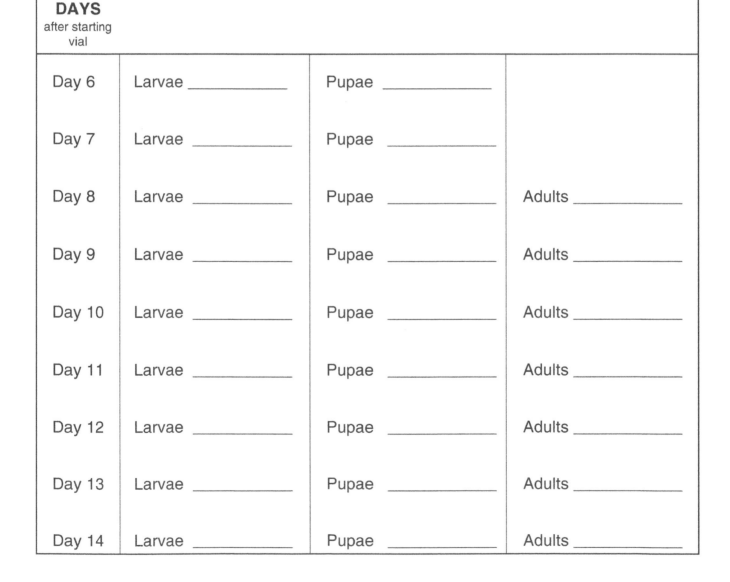

Day 6	Larvae _____	Pupae _____	
Day 7	Larvae _____	Pupae _____	
Day 8	Larvae _____	Pupae _____	Adults _____
Day 9	Larvae _____	Pupae _____	Adults _____
Day 10	Larvae _____	Pupae _____	Adults _____
Day 11	Larvae _____	Pupae _____	Adults _____
Day 12	Larvae _____	Pupae _____	Adults _____
Day 13	Larvae _____	Pupae _____	Adults _____
Day 14	Larvae _____	Pupae _____	Adults _____

Figure 16.3 Blank chart for recording the daily counts of larvae, pupae, and adults.

in as well as the sponge so you'll need to watch the food and maybe spray it with water if it starts to dehydrate.

Add at least eight to ten adult flies that are the same size and look like the same species. The flies will lay eggs on the banana mixture. If you want to see the eggs, put in some black cloth or paper or black banana peel, otherwise they'll blend in with the pale banana. After you've set up the culture and added the adult flies, it requires no maintenance at all. (A "culture" is a container in which organisms are or will be reproducing.) Fruit fly eggs hatch into larvae that grow to a length of about $^1/_8$ inch (3.2 mm). The white larvae will be visible wiggling around just under the surface of the banana mixture about one week after starting the culture. (In nature they burrow into rotten fruit.) About eight or nine days after starting the culture, they'll begin leaving the banana mixture to pupate on the sides of the vials. The pupae look like reddish brown cigars, $^1/_{16}$ to $^1/_8$ inch (1.6–3.2 mm) long. They do not move or eat. New adult flies will begin to emerge from the pupal cases about two weeks after you started the culture and will continue to emerge for about another week. The new flies will begin to lay eggs themselves about two days after emerging. These eggs will also take about two weeks to hatch, go through metamorphosis and reach adulthood. About four and a half weeks after you started the vial you'll begin to get a third generation of adult flies, the grandchildren of those you started with.

When the level of banana mixture gets low in the container, simply start a new vial, adding at least ten adults. This way you can keep your fly population going indefinitely. Each female probably lays a hundred eggs, so there are a lot of adults coming out. If you order fruit flies from a biological supply company, you can get vestigial flies that have no wings. The no-wing condition is a genetic mutation that is passed on to their offspring. They're called vestigials because their wings are vestigial, which means that they are only a vestige of the normal wing. Vestigial flies are much easier to use to feed ant lions and baby praying mantises than are winged flies because they can't escape as easily. You can use winged flies to feed predators, but you may need to anesthetize them by chilling them in the freezer (for one to two minutes, depending on size) so they can't escape as easily.

Field Hunt

You can catch flies by sweeping through tall grass with a net, but they are unlikely to be fruit flies. If you want to hunt for fruit flies instead of setting a trap, hunt around garbage or rotting vegetation. The best way to be sure you're getting fruit flies is to trap them, as described earlier.

Flies at School

Getting Ready

Order vials and banana flakes about a week before you need them. If you have thirty students and want every child to have his or her own vial of flies, you need to start about five vials three weeks in advance to have enough to give each child ten flies. You can instead rely on trapping to start the class's vials, but you'll need to have a lot of traps, or else stagger the students to start their cultures on successive days. If you plan to let the students add the flies to their own vials, you'll need twice as many flies because a lot of them will escape.

Fig, 16.4 Claire records the number of pupae in her vial on her table.

Observations and Activities

Start with adult flies for observation. Bring in a culture vial you've started, or trap some flies. Since the students are used to thinking of flies as houseflies, you may want to ask them to describe how the fruit flies are similar to and different from houseflies. If they ask, you can tell them the definition of a fly: an insect with one pair of functioning wings and a pair of back wings that are just small knobs called halteres. The function of the halteres is to help maintain balance during flight, but you can't see the halteres on fruit flies with the naked eye.

If you want to give the students a chance to closely examine dead flies, you can freeze a few by putting them in a jar in the freezer for ten minutes or more. Ask the students to draw the flies and label their body parts as well as they can (see Figure 16.1). This is a good time to explain the life cycle: egg, larva, pupa, and adult.

In setting up the vials, students can measure out the banana flakes and water and yeast themselves (see Figure 16.2). I like to start out by showing a couple of students how to do it, then having them show the next pair, who shows the next pair, and so on. Make frequent quality control checks to ensure they don't drift too far afield from the original directions, especially with too much yeast. When all the vials or jars have had culture medium (bananas, water, and yeast) added, either you or the students can add the flies. Unless I have a great overabundance of flies, I add the flies myself. If you do let the students do it, spread the flies you have over a great many empty vials before turning them over to the students, so if a vial is dropped you won't lose half your flies.

To add flies from a culture vial to a new vial, it's essential that you first tap the bottom of your culture vial on the tabletop firmly to knock the flies to the bottom of the vial. After this you'll have a few seconds to tap eight to ten flies into the new vial. You need this many to be sure that at least a couple of them are fertile females. Replace both stoppers quickly! It's a good idea to write each child's name on his or her own vial using a small piece of masking tape. Don't let the tape block students' views of the top surface of the banana mixture, or they won't be able to see the larvae.

I give each student a chart for recording daily observations and counts (see Figure 16.3). They can release the parent flies after five or six days because by then the flies have already laid plenty of eggs. With the parent flies gone, the children can take the stoppers out of the vials to look for larvae more carefully. They'll notice about day six or sooner that the banana mixture is more liquid than it was, and they may be able to see the movement of larvae under the surface of the stuff. (The larvae secrete enzymes to soften their food.) The students can see the larvae under the surface by looking

through the side of the vial too. Keep a record of how many students spot larvae for the first time each day.

By day eight most students will have larvae climbing up the sides of the vials to pupate. Even though you've told the students what to expect, many won't understand what's happening. Every morning for a week or so the students will find more of the little reddish brown pupal cases on the sides of the vials, and they'll enjoy counting them (see Figure 16.4). The students can see two respiratory tubes that look like short antennae projecting from one end of each pupa in larger species of fruit flies.

I find that I need to keep going over and over how these white wiggly things and reddish brown cigars relate to the insect life cycle, even if the students can recite the stages of the life cycle perfectly. How do the words *larva*, *pupa*, and *adult* relate to what they're seeing? Remind them that the larvae are eating the banana and yeast mixture. Many children tell me their larvae are eating each other (or fighting or mating or something, all of which don't happen with fly larvae).

One second-grade student I worked with was the son of a teacher and had had his own vial of fruit flies at home for months. He rolled his eyes when I told the class we were going to raise fruit flies. But on the day his larvae first became visible, he didn't know what they were—another reminder that seeing is not necessarily understanding when it comes to critters!

The original adults probably laid eggs every day they were in the vial, so if you left the originals in their vials for a week, they'll get new adults emerging every day over a period of about a week.

Experiments

Remember that the hypotheses I give are just examples. Your hypotheses will be the predictions made by the class or a particular student. Your result for each experiment will be a statement of how your animals reacted to your experimental setup. Your conclusion is a statement of whether your prediction was confirmed or not. For each experiment, adding replicates increases your confidence in the validity of your conclusion, but may be omitted if tedious for young children.

Experiment 1

Question:
How does the degree of rottenness of the fruit affect the flies' interest? Will they come to a jar with fresh fruit or only rotting fruit?

Hypothesis:

If we offer the flies rotting fruit and fresh fruit, they'll like them about the same.

Methods:

Get one grapefruit and three jars. Cut the grapefruit into thirds or quarters. Freeze one piece and leave the other two covered at room temperature. After three days, freeze a second piece that has been at room temperature. After three more days, put the piece that has been at room temperature into a jar and put the two pieces from the freezer into two other jars. Put the three jars outside on their sides, without lids, not more than a foot (.3 m) or so apart. After one day return and put lids on them, slowly, as described earlier. Which one has the most flies? You may get more meaningful results on this one if you do two or three replicates of each setup. You can substitute other fruit, but fruit flies seem to be strongly attracted to grapefruit.

Result:

Your result is a description of what turned up in each jar. You can record your results as a histogram (see Figure 7.4 for an example of a completed histogram). A blank graph for your results is provided in the Appendix.

Conclusion:

The prediction is not confirmed here. In my experience, the older the fruit is, the more the flies like it—until it begins to dry out or turn liquid. Fruit in jars outdoors often dries out before it rots completely. In nature drying is probably less common.

Experiment 2

Question:

Do flies detect their food by sight?

Hypothesis:

Flies use sight to detect their food. If we put food out of sight they won't find it.

Methods:

You need two vials of flies, with approximately equal numbers. You also need two grapefruit halves three to six days old. You're going to put one vial of flies with a grapefruit in the dark and the other vial of flies and grapefruit in the light. You can accomplish this with two small rooms of approximately equal size, like closets or bathrooms, if you have one that can be made completely dark. If so, release one vial of flies in a well-lit room at a good distance (across the room) from the grapefruit. Release the other vial of flies in a completely dark room at a good distance from the grapefruit. Return at intervals of fifteen minutes or so and check the grapefruit in the light room until it has a number of fruit flies on it. Then check the dark room too and count the flies on both grapefruits. Approach the grapefruits slowly or the flies will be startled and leave the fruit. Which has more flies?

Result:

Your result is a statement of your fly counts and other observations. You can record your results as a histogram (see Figure 7.4 for an example of a completed histogram). A blank graph for your results is provided in the Appendix.

Conclusion:

The prediction is not confirmed. Flies locate food mainly by a sense of taste or smell, so the dark should have little effect on their ability to find the food.

Another way to go about this is to offer two items, one of which smells and tastes like food, but neither of which looks like food. Wet one rag with grapefruit juice and another with water and release a vial of flies between them. Do flies go to either one? More flies on the grapefruit-soaked rag suggests that smell or taste plays a part in locating food.

Experiment 3

Question:

Are bananas (or other fruit) and yeast both essential for the flies to live and reproduce? Can they live on banana alone, without the yeast added? Can they live on a mixture of yeast and water alone? What about other variations?

Hypothesis:

Flies can live on fruit alone, but not yeast alone.

Methods:

As a control, set up one vial as usual—one part banana flakes, one part water, and a pinch of yeast.

To a second vial add banana flakes and water as usual, but don't add yeast. In another vial mix yeast and water until it has a pasty consistency. Try other combinations: meat sprinkled with yeast, a vegetable sprinkled with yeast, a rotting wild fruit found outside like a crabapple or persimmon, both with and without added yeast, and a damp piece of bread or cracker sprinkled with yeast. Add eight to ten flies to each vial and wait two weeks. In which vials do the flies reproduce?

Result:

Your result is a statement of your observations of which combinations allow flies to reproduce.

Conclusion:

The prediction or hypothesis is supported somewhat in that fruit to which you have not added yeast may support fruit flies. The flies eat yeast, but fruit left outdoors may very well pick up airborne yeast spores that you don't see. Fruit that has not been outdoors and to which you've not added yeast probably will not support a fly population. The yeast must have something to grow on, so the mixture of yeast and water will not support flies. Yeast grows best on fruit, but it may grow well enough to support the flies on other things as well.

Experiment 4

Question:

Does temperature affect the development time of the flies? How?

Hypothesis:

If we put one vial of flies in a cold place, they will grow more slowly than flies left at room temperature.

Methods:

Set up two or three new vials with banana flakes, water, and yeast as described earlier. Add eight to ten flies to each (be consistent). Put one vial of flies in the refrigerator or outside if it's cool (but not cold enough to freeze them). Leave another at room temperature. If you have a place that's warmer than room temperature, but not more than around 100 degrees F (38° C), put another vial in there. The three vials should be as similar as possible in other regards. In which vial do larvae appear first? Pupae? New adults?

Result:

Your result is a statement of the number of days it takes to reach each stage of development for each vial of flies. Use for comparison the number of days it takes to notice the first larva, the first pupa, then the first new adult in each vial. Since larvae in the banana mixture are sometimes hard to see, you may want to use as your record the first larva you see crawling up on the plastic side of each vial to pupate. There's no mistaking that. You can record your counts on the table in Figure 16.5 and the histogram in Figure 16.6. Figure 7.4 provides an example of a completed histogram.

Conclusion:

The prediction in the hypothesis is confirmed. The warmer it is, the faster the flies will develop, although there is an upper limit. They can tolerate cold almost all the way down to freezing. How could you determine the upper limit of their temperature tolerance? You may get developmental abnormalities at higher temperatures before you get any mortality.

The students can make a graph with days across the horizontal axis and the number of pupae visible on the sides of the plastic vials along the vertical axis. The graph would have two lines, one for the cold group and one for the group at room temperature. The difference in the two lines would show the effect of cold. The graph would look similar to the one in Figure 10.5, which also compares the growth rates of two animals. A blank graph is provided in the Appendix.

Experiment 5

Question:

Does cold affect the number of individuals surviving to pupation?

Hypothesis:

We think that if we put one vial in the refrigerator, it will not make as many new flies as one left at room temperature.

Methods:

Set up two new vials with banana flakes, water, and yeast as described earlier, and add eight to ten flies to each (an equal number to the two vials). Leave one at room temperature; keep one in the refriger-

The Effect of Cold on Fly Development Time

	Cool Flies	Room Temperature Flies	Warm Flies
Number of Days until Larvae Appear			
Number of Days until Pupae Appear			
Number of Days until New Adults Appear			

Figure 16.5 Blank chart for recording results for Experiment 4.

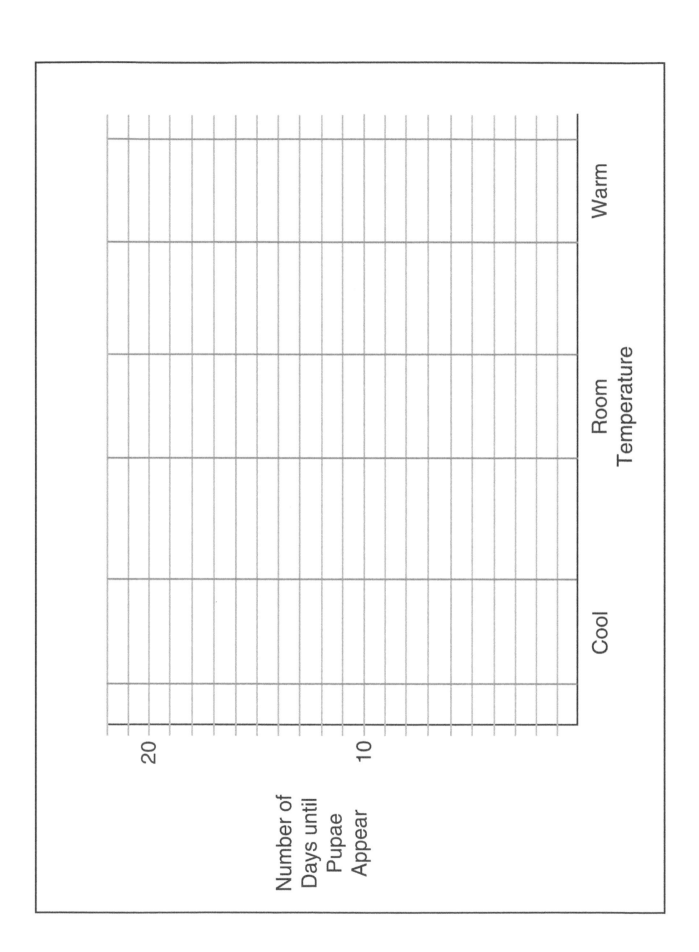

Figure 16.6 *Blank graph for recording results for Experiment 4.*

ator. (You can use the same vials as in Experiment 4.) There is no need, for this experiment, to record any observations about the larvae. After you have begun to get pupae on the sides of one vial, count the number of pupae in that vial every day until you have gotten no new pupae for a period of five days. Use that as a measure of the number of adult flies you are going to get in the first generation of offspring in that vial. Do the same thing when pupae appear in your cold vial. Compare the counts for the two vials. Did you get more pupae in the vial that was kept at room temperature? Using several replicates would add meaning to this experiment, since no two vials are going to produce exactly the same number of adults, even under identical conditions.

Result:

Your result is a statement of your counts of pupae for each vial. You can record your results as a histogram (see Figure 7.4 for an example of a completed histogram). A blank graph for your results is provided in the Appendix.

Conclusion:

The prediction is not confirmed. Temperature should not affect the number of pupae or number of adults, unless the temperature was so close to freezing that some were killed.

Experiment 6

Question:

Does the size of the fly affect the development time?

Hypothesis:

If we start one vial with big fruit flies and one with small fruit flies, the small ones will develop faster.

Methods:

With a fly trap, capture two different sizes of fruit flies. Some fruit flies are at least twice as big as others. Start two vials at the same time, one with large flies and one with smaller flies. In which vial do you see larvae first?

Result:

Your result is a statement of the day you first see larvae, pupae, and adults in each vial.

Conclusion:

The prediction is not confirmed here. Fly species do differ in their development times, but in my experience, this variation does not necessarily correlate with size.

Experiment 7

Question:

Does darkness affect the ability of a fly population to reproduce?

Hypothesis:

If we put one vial of flies in a dark room, they will reproduce normally.

Methods:

Set up two fruit fly vials identically, with banana flakes and yeast. Add eight to ten fruit flies to each (be consistent). Put one in an absolutely dark place—a box on a back shelf in a dark closet, under a dark towel. Leave the other one in a normally lit place. To add confidence in your results, do several replicates of each type of vial. Check both vials or group of vials after two weeks. Is there a difference?

Result:

Your result is a statement of your observations.

Conclusion:

The prediction is not confirmed, probably due to an inability to court successfully in the dark. Some of the male flies' courtship may be tactile (tapping the female with his front legs), but most of it is perceived visually by the female. If she can't see him, she won't be persuaded by his courtship activities to accept him. They probably will not mate, so there will not be offspring in the dark. What would happen if you put the two vials in the light and dark, respectively, after giving the adults a week to mate first and then removing the adults? Would the larvae and the pupae develop in the same way?

Experiment 8

Question:

Are adult fruit flies attracted to light?

Hypothesis:

Adult fruit flies are attracted to light.

Methods:

Put at least twenty fruit flies into a terrarium at least 1 foot (30.4 cm) long with a lid. Put one end of the terrarium near a lamp that's turned on in a room otherwise uniformly lit or darkened. Leave the flies alone for an hour or more, then note their position. As a control, turn the lamp off, wait an hour or so, and note the position of the flies.

Result:

Your result is a statement of your observations.

Conclusion:

The prediction is confirmed. Adult fruit flies are attracted to light and will be clustered at the end of the terrarium near the lamp when it is on.

Experiment 9

Question:

Do fruit flies have a tendency to walk upward, away from gravity?

Hypothesis:

We think fruit flies have a tendency to walk upward, away from gravity.

Methods:

Get two vials of flies and note the position of the flies inside the vials. In both vials most flies will probably be clustered at the top, on the underside of the foam stopper. Turn one of the vials upside down and put it in a dark cabinet. As a control put with it the second vial of flies that you do not turn upside down. Leave the vials in the dark for five minutes. (Put the flies in clean vials first if your banana mixture is very goopy, or the goop may drop in the upside-down vial.) When you open the cabinet, check the flies' position immediately before they have a chance to reorient themselves in response to the light. Did the flies in either vial change position?

Result:

Your result is a statement of your observations regarding the positions of the flies after five minutes in the dark cabinet.

Conclusion:

The prediction is confirmed. The flies in the vial that was turned upside down will have changed their position. They will now be standing on the banana mixture, which is now at the top of the vial. The flies in the other vial will still be on the underside of the stopper. The only difference in the vials is that one was inverted, so the only explanation of the flies' move is a tendency to walk away from gravity.

Chapter 17
Aphids and Their Predators, Ladybug Beetles

Introduction

Aphids are tiny soft-bodied insects that suck plant juices and are major garden pests. Experiments in this chapter address questions about the aphids' feeding and their destruction of garden plants, their reproduction (the females usually reproduce without males), and the relationship between the aphids and ladybug beetles. Ladybugs are voracious predators of aphids and are sold through gardening catalogs for aphid control.

The teacher of one class that kept aphids had this to say about the class's experience: "I ask the children to name all the vowels and no one knows. But I ask them how many aphids were on their plants today and they know that. Everyone remembers because they held them and touched them and owned them."

Aphids are useful particularly for math exercises. The students can make histograms showing how the number of aphids increases each day for the class as a whole or for each student. Older students can calculate averages.

Figure 6 in the Postscript provides a range of questions and answers comparing the life cycles of the organisms discussed in Chapters 16 to 18. A list of the questions represented in Figure 6 is provided in Figure 7, which can be photocopied and handed out for the students to answer.

Materials

1. Sugar snap pea plants (snow pea plants) or lettuce plants planted outside to attract aphids and ladybug beetles. Any other plant that you know will attract aphids will do.

2. Aphids.

3. Ladybug beetles.

4. Small paintbrush for moving aphids. Bristles should be about $1/2$ inch (1.3 cm) long and $1/8$ inch (.3 cm) across.

5. One young sugar snap (snow pea) plant or lettuce plant for each child.

6. One plant container and soil for each child (peat pots, paper cups, or yogurt cups—anything that will hold soil).

7. Fruit fly culture vials and foam stoppers (see Chapter 16 on how to get them) or other small containers with lids.

Background Information

Feeding Aphids

Aphids live on plants, and they feed by piercing stems and leaves with a long, strawlike mouthpart and sucking the plants' sap. Leaves that support large numbers of aphids turn yellow and curl. A heavily infested plant may die.

Aphid Reproduction and Dispersal

Aphids' reproduction is unusual in a couple of different ways. What first caught my attention about them is that most aphid populations are all female and reproduce quite satisfactorily without males. This type of sexless reproduction is called parthenogenesis. All the offspring in such a population are female. Another unusual aspect of aphids' reproduction is that they give birth to live young most of the time instead of laying eggs like most insects. Students can actually see the baby aphids being born if they look carefully. The young aphids mature and begin cranking out their own babies at the ripe old age of

Fig. 17.1 Drawing of an adult ladybug, ³/₁₆ inch (4.8 mm).

one week. One female can have one hundred babies, so the population increases rapidly.

When a plant gets crowded, the aphids need some way to disperse to new plants. Most adult aphids have no wings, but frequent contact with other aphids (crowdedness) signals the females' bodies to produce female offspring that will be winged as adults. These winged female adults fly away singly to establish more populations of unwinged females. (The advantage of the dispersers being all female and parthenogenetic is that it allows them to establish twice as many new colonies as they could if they had to pair up to reproduce.) This cycle of unwinged female populations expanding and finally dispersing via a winged female generation goes on all summer. Only in the fall are some male babies born. When they mature they mate with the females, which then lay fertile eggs instead of bearing live young. The eggs overwinter and hatch in the spring into another parthenogenetic all-female population.

Why Mate at All?

If reproduction without sex allows them to start more new colonies, why mate in the fall or ever? The sexual reproduction (mating) in the fall provides some genetic variability in the offspring. Siblings from sexual reproduction are not all identical, just as most human siblings are not identical. But sisters of a parthenogenetic or female-only reproduction are all clones of their mother, all genetically identical.

When all individuals in a population are genetically identical, the population is more vulnerable to environmental stress. Genetic variability in a population increases the likelihood that some individual will possess the traits necessary to weather any particular crisis.

Population Growth

Aphids are a useful tool in talking about rapid population growth. The human population is growing much too fast as is, and we've already exceeded the carrying capacity of our planet. That is, our consumption of resources exceeds the ability of the planet to renew, replace, or regrow them. So, we're running out! People around the globe are already facing shortages of vital resources such as fuels and arable land. These issues will become much more critical as your students reach adulthood and we add an additional two billion people. To see estimates and graphs of human population growth worldwide, check out the interactive webpages of International Data Base.

The same population surge would happen to any insect or animal population that was left unchecked. A toad may lay one hundred or two hundred eggs at once. What if each one of the eggs grew up and became the parent of one hundred more eggs? And each of those eggs grew up and did the same? In three years or less, there'd be a million (100 x 100 x 100 = 1,000,000) toads instead of two. But in reality, only a tiny fraction of toads in nature survive to adulthood, so that doesn't happen.

What about aphids? Why is the world not covered with aphids? Because they have lots of predators. Ladybug beetles are their primary predator, which is why they are sometimes sold for controlling aphids. This lesson is an opportunity to show students (1) that any unchecked population will increase rapidly, and (2) that predation can provide population control.

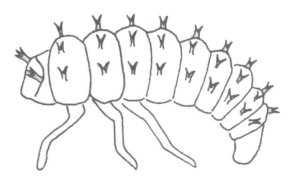

Fig. 17.2 Drawing of a ladybug larva.

Predation is one of the primary factors that can limit population growth in nature. Competition is another. What can limit human population growth? It's complicated at this point. But we know a few things. The education of girls, empowerment of women, and economic opportunities for women in developing countries have profound effects on worldwide population growth. Women who have a voice in controlling family size and who have options in life other than bearing large families generally choose to limit their family size. If you want to invest in a future viable planet, those are good places to start.

Fig. 17.3 Drawing of an aphid.

Ladybug Beetles

Ladybug beetle adults and larvae are predators of aphids. There are many different species of ladybugs, and the adult coloration varies. Some are black with orange or yellow spots (see Figure 17.1). Ladybugs lay eggs on plants that harbor aphids. The eggs hatch after several days into tiny ($\frac{1}{16}$ to $\frac{1}{8}$ inch [1.6–3.2 mm]) black caterpillar-like larvae (see Figure 17.2). Larvae of some ladybug species can eat up to five hundred aphids a day! Most larvae molt (shed their exoskeletons) three times as they grow. When the larvae are about $\frac{1}{4}$ to $\frac{3}{8}$ inch (.6–1 cm) long, they are ready to pupate. They affix themselves to a surface and remain immobile. The skin of the larva splits and the new skin underneath hardens as it becomes a pupa. Inside the pupal skin an adult is forming. It

doesn't eat during this process of metamorphosis. After about five days, an adult will emerge from a split in the hard skin of the pupa. The characteristic spots of the adult ladybug appear gradually as the wings harden. The adult, like the larva, walks around the aphid-infested plant, eating aphids. It never touches the aphids with its legs, but just walks right up to an aphid, grabs it with its tiny jaws, and munches.

The most outstanding features of aphids to young children are these:

1. The number of aphids on a plant can increase from one to fifteen overnight.
2. Aphids give birth to live young.
3. A dense population of aphids can sicken a plant.
4. Some aphids are winged and some are not.

How to Get and Keep Aphids and Ladybugs

Aphids

I don't know of anywhere to order live aphids, but they're easy enough to come by. I've never had a garden that wasn't plagued by aphids sooner or later. They're always on sugar snap pea plants and always on leaf lettuce. In the South sugar snaps can be planted in mid-February, or even earlier, and sprout in March. Lettuce is planted in early- to mid-March. Aphids appear on both in April (later in the North). You may need to start your plants indoors in the North and set them out when it's warm enough to get aphids before summer vacation. Ask someone at a local gardening supply store how early you can plant.

When the aphids first appear there are just a few. Look on the undersides of leaves and where there is new growth. Aphids come in lots of colors, but most I've seen are green or red. An adult is usually the size of the head of a pin, sometimes twice that size. They are pear-shaped (the narrow end is the head) and very soft-bodied insects (see Figure 17.3). Their legs are so thin they are barely visible. Aphid eyes are tiny black specks, which I didn't notice until a first grader pointed them out to me.

To maintain aphids, just keep their plant alive, leave them on the plant, and don't use any pesticides near them. Since they prefer tender growth and buds, don't prune back all the new shoots. There are many aphid species, and each species is restricted biologically to feeding on only one or a few particular types of plants. Its innately preferred plant species is called its host plant.

Aphids will die or leave if placed on plants other than their host species.

When you need to collect aphids to put on students' plants or to feed ladybug larvae, you can collect them from a plant with a tiny paintbrush and a vial if you've gotten some vials for baby mantises or fruit flies (see Chapters 15 and 16). Put the open vial directly under the aphid. Then gently sweep the aphid into the vial and put the foam stopper in. You can also pick up an aphid (or any other tiny insect) with a paintbrush by touching the blunt end of the brush gently to the aphid. Its legs get tangled in the bristles and you can lift it easily and without damage. Then gently tap the paintbrush against the edge of the vial; the aphid falls into the vial.

Ladybugs

If you want to do the experiments involving predation (Experiments 5 and 6), you'll need ladybugs. I've always been able to find at least a couple with the help of a class of students. Even first graders probably know what a ladybug looks like. If you take a class into a natural area for a half hour in warm weather, chances are someone will find one. Have them look at home and at recess for a week. They can also be ordered from a biological supply company, although garden supply sources are usually cheaper and send huge quantities (see Appendix for both). I keep ladybug beetles in vials with the wet end of a cotton swab in the vial for moisture held in place by the foam stopper (see Figure 17.4). A jar with a bit of moistened paper towel or cotton will do.

If you find a beetle it may lay eggs in the container. They wait about a week after mating before egg laying. Another way to get eggs is to find them already laid. If you carefully examine aphid-infested plants that have been outside, there's a good chance you'll find some ladybug eggs. The eggs look like a tiny cluster of short, thick, yellow cigars (white when about to hatch), all on end, with their sides stuck together more or less (see Figure 17.5). Each egg is about $1/16$ inch (1.6 mm) long. The eggs of other beetles that may be attacking the plant instead of the aphids look similar, but chances are if you find eggs like these on an aphid-infested plant they're ladybug eggs. Check every day for a week or two. If you find them, leave them on the leaf. Put the leaf in an escape-proof container, such as a vial or small jar. The eggs are so tiny the air in the jar will last for a long time. After they hatch, the larvae stay clustered for a few hours, then disperse rapidly looking for aphids to eat. I usually let all but a few go and put the remaining

ones in separate vials with one end of a cotton swab, as described for the adults. Ask the students how many want a larva for Experiment 5 or 6. Keep that many, if you have enough, or keep enough to do those experiments as a class.

If the students put their plants outside for some sunshine, they may get some ladybug eggs that way. It happened the first time I shared aphids with a class. A little girl named Carmen discovered eggs on her plant by herself after we brought the plants in, and was delighted when she found out what they were. The next morning they hatched. Carmen was thrilled to be the owner of our only ladybug larvae.

Ladybug larvae and adults need aphids to eat. I have offered alternative diets but have not found any to be accepted. Although they can eat huge quantities of aphids, only a few aphids a day will keep them going. Like all cold-blooded (poikilothermic) animals, ladybugs

Fig. 17.4 Drawing of a fruit fly vial with a cotton swab for moisture for ladybug larvae and aphids.

have a lot of flexibility in the quantity of food consumed. You can collect aphids from a plant with a paintbrush, as described earlier, and then tap them into the ladybugs' vials. I feed ladybug larvae and adults by collecting aphid-covered buds from a thistle weed behind my house and putting a whole bud in with each predator.

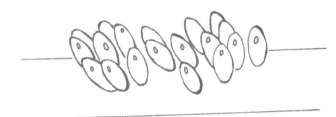

Fig. 17.5 Ladybug eggs, $^1/_{16}$ inch (1.6 mm).

Field Hunt

A garden is a good place to look for aphids, but they can be found on many other plants as well—both cultivated and wild. Show online photographs of aphids to the students before you hunt. Gardening pest books or websites have pictures too, since aphids are common pests. Tell the students to check stems and leaves, especially the undersides and new growth.

My son, Alan (then three years old), tried to collect aphids from our garden on his own after watching me do it. He took the foam stopper out of the vial and began sweeping them jerkily into the vial with his paintbrush. He was knocking most of them onto the ground because the aphid's only defense is to simply let go and drop

Fig. 17.6 Tyewanda tamping down the soil after planting her sugar-snap pea plant.

to the ground when disturbed. Another defense is that some aphids lure ants to defend them by feeding the ants excess sugar the aphids have sucked from the plants. The excess is secreted by the aphids as a liquid called honeydew, edible to the ants.

After a demonstration of how to pick up an aphid with a paintbrush, Alan tried again and got several into his vial.

Ladybugs are found almost anywhere outdoors. Have the students search plants and the ground. Ladybugs are found most easily perhaps by not looking, and being patient and waiting until one finds you.

Aphids and Ladybugs at School

Getting Ready

A month or maybe more before you plan to begin the lesson, plant some sugar snaps or whatever plants you plan to use later in the classroom. Plant in pots and transfer them outside as weather permits. The purpose of this is to attract aphids so you'll have some to transfer to the students' individual plants in the classroom. After you've found aphids on your outdoor plants, you can start growing the students' plants in cups.

Observations and Activities

A good place to start involving the students is in planting their own host plants for their future aphids (see Figure 17.6). They can use paper cups, peat pots, or plastic yogurt cups. I generally use small milk cartons that the students collect themselves from lunches. Either potting soil or good soil from outside will do. The sugar-snap seeds will germinate faster if you soak them for twenty-four hours or so before planting. Label the cups with the children's names and, if possible, let each fill his or her own cup with soil, leaving about 1 inch (2.5 cm) of space at the top. Let each plant a couple of seeds. (Even a toddler can do this.) If the soil is kept moist, the seeds should sprout within a few days. When a plant has at least two unfolded leaves, it's big enough for aphids. Most plants will be 3 to 4 inches (7.6–10.2 cm) tall about a week after planting. Don't let the plants get much bigger before putting the aphids on them, or the students may have trouble finding the aphids later.

The easiest way to introduce aphids to the students' plants is to use the paintbrush to gather a vial full of aphids from your source plant. Then, again using the paintbrush, put an aphid from the vial onto each child's plant. Or you can let the students transfer the aphids to their own plants. I've found third graders can do it easily, if they do it one at a time. I show the first person how to do it, then he or she shows the second person, and so on (see Figure 17.7).

Fig. 17.7 Shawnta and Valerie are trying to get an aphid onto Shawnta's plant.

Tell the students at this point that the aphids are sap-sucking insects with strawlike mouthparts. You may want to stop there and see how much the students discover on their own.

The day after you introduce the aphids to the students' plants, have each child count the number of aphids on his or her own plant. Although each started

Fig. 17.8 Mena counts the many aphids on her plant the day after putting her single aphid on it.

with only one, most will have several more now. Most of the new ones will be babies (smaller than the parent). Some will be newly arrived winged aphids that flew in from outdoors.

They'll ask questions about the new arrivals. If the students are old enough to know that most animals mate to have babies, they may wonder how there can be babies with only one parent. This is a good time to give them a little information on the aphid life cycle, or at least the parthenogenetic part. Help them to notice that the aphids are giving birth to live young (you can see this happening).

Have them count the baby aphids for a week or more (see photo in Figure 17.8). If the students are old enough, each child can make a histogram of his or her own results on Figures 17.9 and 17.10, with the sequence of days on the horizontal axis and the number of babies (per day or cumulative) on the vertical axis. The cumulative count will show the rapid increase in the population.

Experiments

Remember that the hypotheses I give are just examples. Your hypotheses will be the predictions made by the class or a particular student. Your result for each experiment will be a statement of how your animals reacted to your experimental setup. Your conclusion is a statement of whether your prediction was confirmed or not. For each experiment, adding replicates increases your confidence in the validity of your conclusion, but they may be omitted if tedious for young children.

Experiment 1

Since aphids are considered garden pests, one obvious question is this:

Question:
What happens to the plant when it gets crowded with aphids?

Hypothesis:
If the plant gets crowded with aphids, it will get sick.

Methods:
Allow aphids to multiply unchecked on one plant. On another plant keep aphid numbers low as a control (for comparison). Rate the condition of the plants every day.

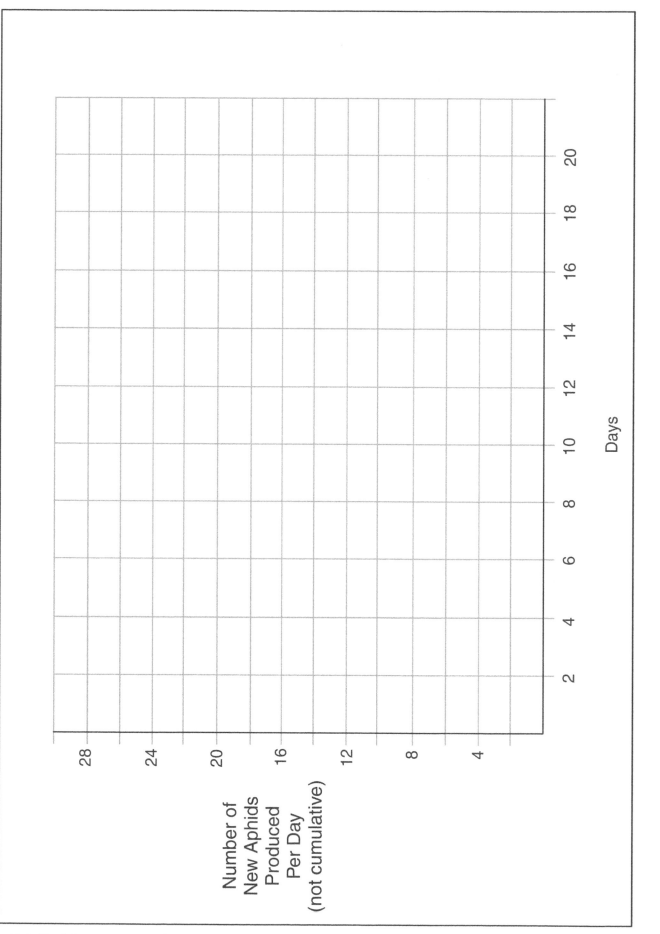

Figure 17.9 Blank histogram to record aphid tallies per day on.

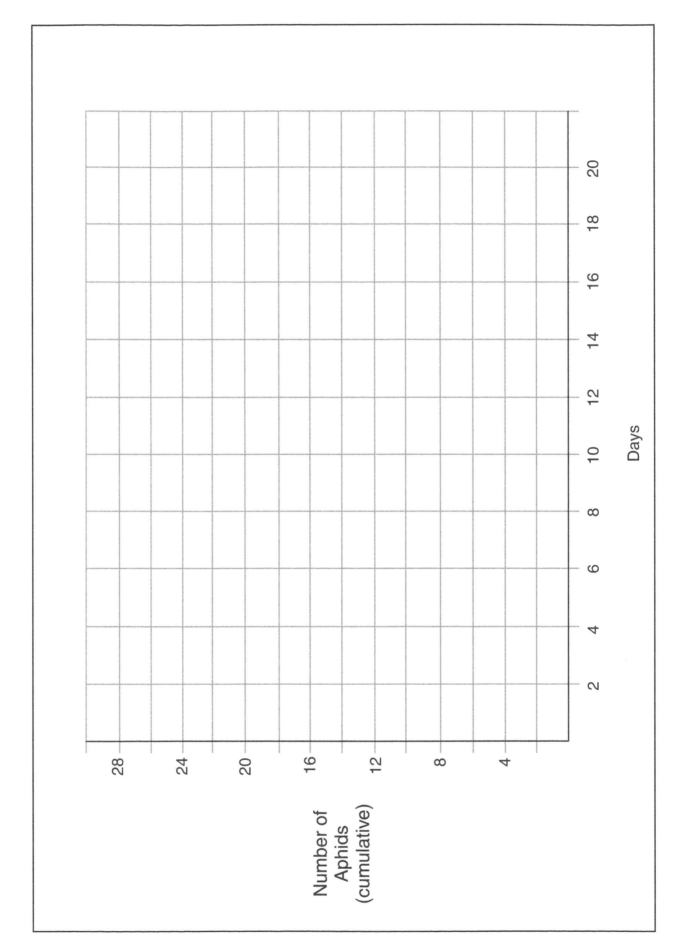

Figure 17.10 Blank histogram to record cumulative aphid tallies on.

Result:

Your result is the relative condition of the two plants after a period of several days.

Conclusion:

A very crowded plant turns yellow, gets curled leaves, and eventually dies.

Experiment 2

All animals have the problem of dispersal, as do humans. If a farmer has four children, and they have three or four children each, the offspring can't all inherit their parents' farm. How do animals handle this problem? If all the seeds from a maple tree sprouted under the tree, the competition for light, water, and soil nutrients would be so fierce that very few or none would survive. So the seeds are equipped with wings to disperse via the wind.

All living things, plant and animal, have some method whereby the young can spread out, or disperse, to avoid interfering with one another. Many seeds are dispersed by animals that eat them and later eject them unharmed in their feces. This is why fruits have fleshy edible parts. Other seeds have spurs that cling to animals' fur or human clothing and are thereby dispersed. Many animals simply wander or fly away when grown.

What do aphids do when they get crowded? Do they disperse? How? An aphid on the end of a stem has a long way to wander to a new plant. You might pose this question to the students. What do they think? If they're stumped, get them to think about how other insects, such as flies, beetles, and dragonflies, disperse. Most fly away. They may or may not come up with suggestions, or hypotheses, but lead the conversation in such a way that someone (maybe you) proposes that you let the plants get too crowded and see what happens.

If your plant starts to decline from lack of sunshine, the aphids may die before they get crowded. You may need to set it outside for a few hours a day if you don't have a sunny window.

Question:

Will aphids disperse when they get too crowded?

Hypothesis:

If we let one plant get too crowded, the aphids will somehow leave.

Methods:

Let one or more plants get too crowded. Add new adults if you want to speed things up. Keep an uncrowded plant as your control (for comparison). Watch the aphids every day. This is the same setup as Experiment 1, but you're watching the aphids instead of the plant this time.

Result:

Your result is a description of specifically what the students observed about the aphids on the crowded plant and on the uncrowded plant. Your anticipated result is that the females on the crowded plant will begin giving birth to young that will be winged as adults and will fly away. The aphids on the uncrowded plant will not have young that grow up to be winged.

To get this result, both your plants must get adequate sunlight and water so that the health of the plant is good other than the effects of crowding. Aphids don't reproduce enough on a sickly plant for it to work.

Conclusion:

Aphids do have a way of dispersing when crowded.

The stimulus that causes this change to winged offspring may be an increased number of contacts between the mother and other aphids, or it may be some alteration in quantity or quality of sap.

If you can't get the answer by this approach, try it from the other end. That is, put a fresh plant outside and see how new aphids get to it. You may need to put out several or wait for a while. If it's warm outside, eventually you'll get a winged aphid. Have the students make hypotheses about how they think the aphid will get there. If they've made a prediction, the observation of what really happens will make much more of an impression on them.

Experiment 3

The sap of a plant travels in vessels similar in principle to our blood vessels, although there is no heart. If you cut off a stem, you stop the flow of sap in that stem. What happens to the aphids if you cut off the stem and the flow of sap?

Question:

What will aphids do if the stem is clipped?

Hypothesis:

If the stem is clipped, the aphids will starve.

Methods:

Clip a stem with aphids on it and keep an eye on the aphids over the next couple of days. Watch the aphids on a still attached stem for comparison, as your control.

Result:

Your result is a description of the aphids' behavior on the cut stem and on the attached stem.

Conclusion:

The aphids will leave the cut stem or die. If another stem is available, they find it and climb it from the soil up. Aphids should stay on an unclipped stem, provided the plant is healthy. A clipped stem can't sustain aphids.

Experiment 4

A first-grader, Courtney, became interested in transferring her aphids to other types of plants and making hypotheses about whether the plants would be accepted.

Question:

Will my aphid stay on grass from the lawn?

Hypothesis:

If I transfer an aphid to a grass plant, it will stay.

Methods:

Put an aphid or aphids on the grass. The grass must be healthy and rooted in soil. Transfer an aphid to another pea plant simultaneously as a control. If the aphids leave both plants, then being handled may have caused the exit. If only the aphid on the grass leaves, then the grass is probably not a host plant of that aphid species.

Result:

Your result is a description of the response of both transferred aphids. Does each stay or leave?

Conclusion:

It depends on the species of aphid and plant.

Experiment 5

You may want to begin this experiment without telling the students that the ladybugs will eat aphids.

Question:

Will ladybugs eat aphids? Can ladybugs control aphid population growth?

Hypothesis:

If we put ladybugs on our plants, they'll get rid of some of our aphids.

Methods:

Give each student who wants one either an adult or a larval ladybug for their own snow pea or lettuce plant. They may need help getting the predators onto a stem with their aphids. Show one student how to do it, then he or she shows a second student, and so on. This way there are always only two at the table. Let them take the plants back to their desks and watch what happens right away because sometimes the ladybug will leave. After they've seen the ladybug in action, they can then put the plant back on a windowsill to get some sun. Keep another similarly infested plant free of ladybugs for comparison as a control. Wait a few days. Compare the health of infested and uninfested plants. Which plant seems healthier?

If the students don't see any predation occurring on the plant, you can enclose the ladybug and aphids in a vial together and watch (without any plant parts). In this event, the students will have to speculate about the outcome of the second question. You can also enclose the end of a live stem, with aphids and a ladybug, in a vial with a foam stopper. The stopper keeps the insects in without crushing the stem, which is still attached to the rooted plant. This works well to contain wandering ladybugs, but children usually wind up breaking the fragile stem by careless handling of the vial.

Result:

Your result will be the students' observations about the effect of ladybugs on the aphid populations.

Conclusion:

Ladybugs eat aphids voraciously. If the plants are heavily infested in the beginning, the one without ladybugs will eventually die. The ladybugs should reduce the aphid population on the other plant, thereby saving the plant. If the ladybugs won't stay on the plant, they'll certainly eat aphids in a container. Because ladybugs are effective in controlling aphids, they are a help to gardeners.

Experiment 6

You may want to talk about gardening without toxic pesticides here. Do the students think that ladybugs can do as good a job in protecting the plant as an insecticide could do?

Question:

Which is more effective in controlling aphids, ladybugs or chemicals?

Hypothesis:

Ladybugs (or chemicals) are more effective.

Methods:

Use a spray bottle to spray an infested plant with soapy water (one application). To another similarly infested plant, add a few ladybugs. After a few days, which has more aphids?

Result:

Your result is a description of what happened when ladybugs or other agents were applied to the aphid populations. The result depends in part on how thoroughly you spray and on whether or not the ladybugs stay on the uninfested plant.

Conclusion:

Ladybugs can be as effective as some sprays in controlling aphids.

What advantages do ladybugs have over toxic chemical sprays? What about soapy water and alcohol? (Alcohol and soapy water are not toxic as long as the vegetables are washed, while the chemicals sold as pesticides can persist even through washing because they are often incorporated into the flesh of the vegetable.) Chemicals sold as pesticides are probably the most effective aphid killers, but you have to weigh their effectiveness against their toxicity to humans and to the other living things outside. The students can weigh the advantages and disadvantages of each method of aphid control and make recommendations.

Chapter 18
Insect Metamorphosis on a Macro Scale: Mealworms to Beetles

Introduction

Most of this chapter focuses on *Tenebrio* beetles, with a quick glimpse at flying beetles. The outstanding feature of *Tenebrio* beetles is their illustration of the insect life cycle on a much larger scale than what we see with fruit flies. The *Tenebrio* larvae, commonly called mealworms, grow to about 1 inch (2.5 cm) long, which is ten times longer than fruit fly larvae. All three life stages of the beetle—larvae, pupae, and adult—can be handled and inspected freely (see Figure 18.1).

These beetles, or their larvae (mealworms), are often available at pet stores as food for amphibians and larvae. Full-grown praying mantises (see Chapter 14) enjoy the adult beetles and eat mealworms like corn on the cob, grasping one end in each front leg.

The activities and experiments with *Tenebrio* include exploratory behavior, habitat choices, growth rates, food choices, and more. I've included a list of questions for flying beetles, which can be caught at certain times of year when they fly to a porch light or lamp outside.

Figure 6 in the Postscript provides a range of questions and answers comparing the life cycles of the organisms discussed in Chapters 16 to 18. A list of the questions represented in Figure 6 is provided in Figure 7, which can be photocopied and handed out for the students to answer.

Materials

For *Tenebrio*:

1. A larva for each student or group to watch through metamorphosis, plus twenty or so for various experiments. Available from pet stores or a biological supply company (see Appendix).

2. A dishpan or other container of similar size with no lid. This can be ordered with the mealworms, or bought at a store.

3. Oat bran or wheat bran to fill the container half full. Oat bran is preferred. Bran without additives can be found in the hot cereal section of most grocery stores, at health food stores, or ordered with the mealworms.

4. A raw potato.

5. Strips of newspaper.

6. A petri dish $3^1/_2$ inches (9 cm) in diameter for each student or group (available at a science hobby store or biological supply company, see Appendix).

7. Refrigerator or other cool place for Experiments 1 and 9.

8. Construction paper for Experiments 3, 4, and 5.

9. A hot water bottle, bowl of ice, and baking tray for Experiment 6.

10. Alternate food and a box for Experiment 8.

11. A box for Experiment 11.

Background Information

Tenebrio is the most useful beetle for classrooms that I know of, and it is certainly an easy one to keep going generation after generation. Each student can keep his or her own larva to raise to adulthood in a petri dish with a small amount of bran and a cube of raw potato. The wormlike larvae are soft, light brown, and easily viewed and handled at their maximum length of about 1 inch (2.5 cm). They have six short legs as do all insects. The larvae don't have three body parts, but the adult beetles do. Students will want to know if their larvae have mouths and eyes, which they do. The students will also want to know the gender of their larvae, but you can't tell that until you watch them mate as adults.

The larvae are very lethargic. When they do move, it is slowly and clumsily. Their exoskeletons are shed or molted about five to seven times as they grow. The duration of the larval period varies depending on temperature and food availability, but lasts about ten weeks under ideal conditions. It can last several weeks longer.

When a larva molts for the last time, the soft new exoskeleton underneath has the shape of a pupa. This change from larva to pupa is its first metamorphosis, which means "changing form" (see Figure 18.3).

Fig. 18.1 Malcolm expertly displays his two adult Tenebrio *beetles.*

The pupae don't move much unless handled, and they do not eat. Two to three weeks after pupation, an adult beetle, about $5/8$ inch (1.5 cm) long, emerges from the pupal exoskeleton (see Figure 18.4). The adults are almost white right after metamorphosis, but soon turn dark brown. I've never seen one fly, but they crawl constantly.

The adults mate a few days after emergence. At this point you can tell males from females because the male crawls on the female's back. You can see the tip of his abdomen curved downward to reach the tip of her abdomen, to transfer the sperm to her.

The female beetles lay eggs about seven to ten days after emergence from the pupal exoskeleton. The eggs hatch about fourteen days later into tiny larvae, which are not really visible in the bran for several weeks.

How to Get and Keep Beetles

Tenebrio *(Mealworms)*

If you order *Tenebrio* beetles from a biological supply company, they come with instructions and a supply of bran for food. I keep mine in a plastic dish tub over half full of bran. They need a fresh slice of raw potato on the surface of the bran every three days or so for moisture. Other than that I haven't touched my culture in several months, and it is thriving with adults, larvae, and pupae. My kids play in it sometimes, fishing out the beetles. Once in a while I take out a larva or beetle to feed a praying mantis or some other predator. The adults don't live very long, so there are dead adults in the culture. I never remove them, and it doesn't seem to make any difference. They just dry out.

When the bran under the surface begins to get powdery, or if the culture seems too crowded, start a new one by transferring some larvae and/or pupae to a new pan of fresh bran.

Students will want to know where the beetles and larvae are found in nature. *Tenebrio* flourish in places where large quantities of grains are stored, such as on farms. They are seldom found in kitchens. In nature, they live under the bark of dead trees. I don't know positively how to tell *Tenebrio* apart from the many other kinds of beetles that live under the bark of dead trees, so I either buy them at pet stores or order them online. They're cheaper at pet stores, which sell them as food for various pets.

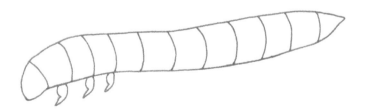

*Fig. 18.2 Drawing of a mealworm (*Tenebrio *larva), ¾ inch (1.9 cm) in length.*

Beetles at School

Getting Started

If you want students to watch *Tenebrio* beetles go through complete metamorphosis, I suggest starting with small larvae. The students can watch the larvae grow to full size before pupating. The population you get from

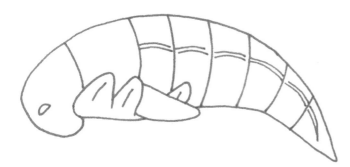

Fig. 18.3 Drawing of a Tenebrio *pupa, approximately ½ inch (1.2 cm) in length.*

a biological supply company is usually an assortment of adults, larvae, and pupae. You may want to ask how many small larvae you can expect. If you want a small larva for each child, you may have to order your beetles several months in advance and let them reproduce. Or you can always start with large larvae. The smallest larvae I'm able to find in my culture are about the length of a child's fingernail. Larvae smaller than that are too hard to see in the bran.

Each student will need for his or her mealworm a petri dish with a lid or other similar container like a baby food jar, 1 tablespoon (15 ml) of bran, and a small chunk or slice of potato for moisture. The students enjoy scooping the bran into their own dishes and selecting their own mealworms. Use a grease pencil to write names on the petri dishes (plastic petri dishes crack with too much pressure).

Observations and Activities

Tenebrio

After setting up a dish for each child, have the students measure the mealworm every other day or so and record the date and length in a journal. Have them record the dates their larvae metamorphose into pupae.

After all have metamorphosed into pupae, make a list of each larva's final length measurement before pupation. Were they all about the same length before pupating? How much variation was there, if any? Can a larva be induced to pupate at a somewhat smaller size than average by removing it from its food source?

Have the students record the dates their pupae metamorphose into adults and calculate the duration of the pupal stage. Make a class list of these calculations. Were the pupal stages of the same duration? How much variation was there, if any? If there was variation, help students calculate an average and state a range; for exam-

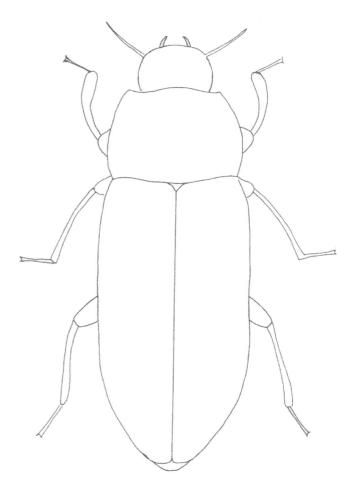

Fig. 18.4 Drawing of an adult Tenebrio *beetle.*

ple, "The longest pupal stage was Kelly's, with x days. The shortest was Alicia's with y days."

Exploratory Behavior of *Tenebrio*

Here are some questions to help students notice the exploratory behavior of an adult beetle. Have each student put an adult beetle on top of his or her bare desk.

- What does the beetle do? Sit still? Move around constantly?

- Walk around the perimeter of the desk? Or walk in a straight line and fall off?

- How does a beetle react to an obstacle (petroleum jelly, a pencil laid in front of it, water, or a child's hand)? Does it turn around and take a new direction? Climb over or through? Try to go around?

- Compare its behavior to that of other animals in this book in the same setting, like a roly-poly or a millipede.

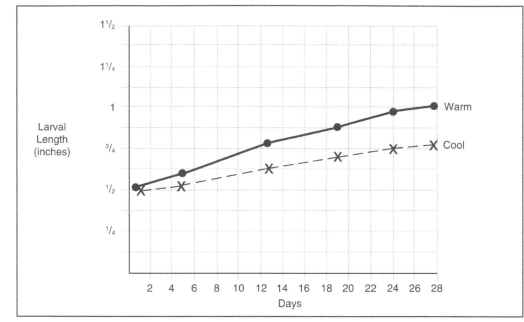

Figure 18.5 Line graph showing larval growth rates in two different temperature environments.

How does the adult beetle behave when exploring a new place with food? Get a square box about 1 foot (30.4 cm) in diameter. Put a pile of wheat or oat bran in each corner. Put a beetle or beetles (adults or larvae) into the middle of the box.

- Do the beetles go directly to a pile? Or do they seem to find it by accident?

- Do they stay in the pile once they find it?

- Does one beetle, when put in the box alone several times, go over and over to the same pile?

- Does a second or third beetle go to the same pile as the first one?

- Do they prefer a pile with a potato slice to a pile without one? (Give them a day or two to make this choice.)

Experiments

Remember that the hypotheses I give are just examples. Your hypotheses will be the predictions made by the class or a particular student. Your result for each experiment will be a statement of how your animals reacted to your experimental setup. Your conclusion is a statement of whether your prediction was confirmed or not. For each experiment, adding replicates increases your confidence in the validity of your conclusion, but may be omitted if tedious for young children.

Experiment 1

Question:
How does temperature affect the speed of development of *Tenebrio* larvae?

Hypothesis:
We think mealworms kept in the cold won't grow at all.

Methods:
Keep one group of larvae at room temperature. Keep another group in a cooler environment; a refrigerator is ideal. An air-conditioned room and a non-air-conditioned room might make a difference. Or you can put one group of larvae outside if it's cool, as long as the temperature is above freezing.

Result:
Your result is a statement of your measurements on particular days and relevant observations.

Conclusion:
The prediction is not confirmed. The larvae develop more slowly at cooler temperatures (see Figure 18.5). They may also molt fewer times and pupate at a smaller size than larvae under ideal conditions. You can make a graph like Figure 18.5 showing the difference in the two growth rates, using the blank graph in Figure 18.6.

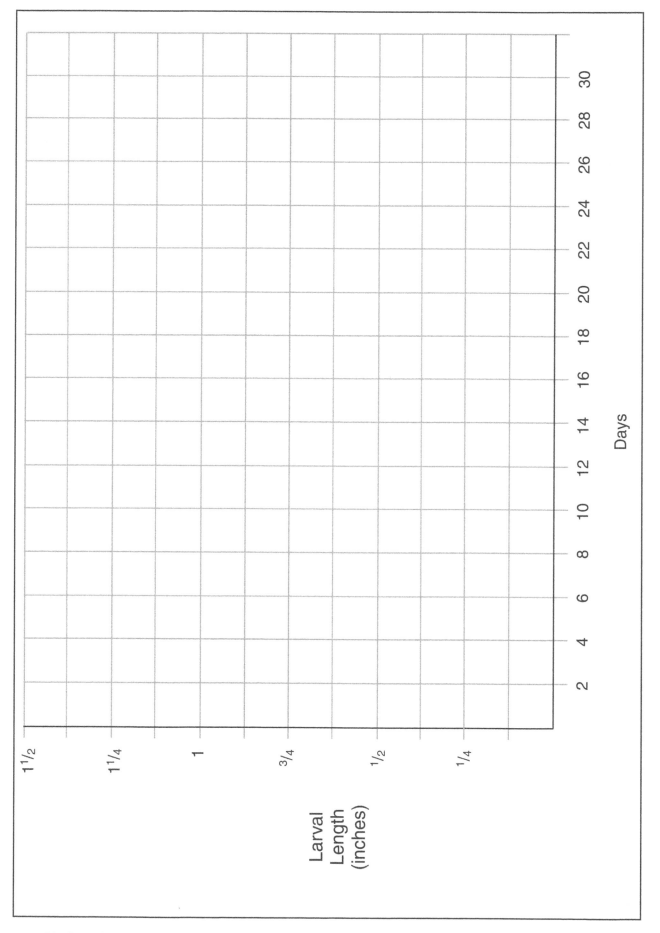

Figure 18.6 Blank graph to plot your measurements of larval growth rates in two different temperature environments.

Experiment 2

Question:

How does food supply affect the speed of development of *Tenebrio* larvae?

Hypothesis:

We think larvae that don't get enough to eat will get sick.

Methods:

Select two groups of similar-sized larvae. Provide one group of larvae with more than enough bran at all times. Feed the other group intermittently. For example, surround them with bran every other day or every third day. On intervening days, remove them from the bran entirely. Handle them gently to make sure that injury is not the cause of any observed slowing of growth. To be certain that handling is not the cause of any observed difference between the two groups, you'd need to handle the larvae remaining in the bran also. Provide moisture (a fresh potato chunk) for all.

Result:

Your result is a statement of your measurements on particular days and any relevant observations. You can plot your measurements for these two groups on a graph (see Figure 18.5 for an example). A blank graph for your measurements is provided in the Appendix.

Conclusion:

The prediction is not confirmed. Those fed intermittently will grow and develop more slowly than those fed continuously.

For each of the following four experiments, you should have at least ten replicates because not every larva and beetle will make the same choice. The choice of the majority will determine your conclusion.

For Experiments 3 to 5, each beetle or larva should be in a dish alone, so they won't influence each other's behavior. If each child has his or her own petri dish and larva or beetle, then each child's dish will be one replicate. You can record your results for Experiments 3 to 7 on a table similar to the one shown in Figures 7.7 and 7.8.

Experiment 3

Question:

Do *Tenebrio* (beetle or larvae) prefer dampness or dryness?

Hypothesis:

Beetles and larvae will choose dampness over dryness.

Methods:

Provide a damp surface and a dry surface in each container (petri dish or other). The texture should be the same on the two sides. I use two half-circles of paper in a petri dish, one damp and one dry. With paper, you have to keep the two sides apart a little so the dry paper won't wick water from the other. Tape down the edges of the paper so the larvae or beetles can't crawl under it. You can also use wet and dry sand. In this case make a foil barrier between wet and dry (no higher than the surface of the sand).

Put a beetle in the middle of each container, and after a half hour record their positions. After you've done beetles, try larvae.

Result:

Your result is a statement of the number of beetles on the damp side and the number on the dry side, as well as the number of larvae choosing each side. You can record your results as a histogram (see Figure 7.4 for an example of a completed histogram).

Conclusion:

The prediction is not confirmed. In my experience, adult beetles prefer the dry side. They may initially suck water from the wet side if they are dehydrated, but most will eventually wind up on the dry side. On the other hand, the larvae do not show a strong preference (in my experience) for either the wet side or the dry side.

Experiment 4

Question:

Do *Tenebrio* (beetles or larvae) prefer light or dark?

Hypothesis:

Beetles and larvae prefer light to dark.

Methods:

Cover half of the lid of each petri dish with black construction paper, so that one-half of the dish is in deep shadow (see Figure 18.7). Leave the floor of the dish bare. Put one beetle in the middle of each dish and put the half-covered lid in place. Record the location of the beetles after a half hour or longer. Then repeat the procedure with larvae.

Fig. 18.7 Craig assesses his beetle's choice choice of light versus dark.

Result:

Your result is a statement of the number of beetles under the black construction paper and the number in the light, as well as the number of larvae choosing each side. You can record your results as a histogram (see Figure 7.4 for an example of a completed histogram).

Conclusion:

The prediction is not confirmed. A majority of adult beetles choose the dark side. In my experience the larvae do not show a strong preference for either side.

Experiment 5

Question:

Do *Tenebrio* (beetles or larvae) prefer a dark or a light substrate?

Hypothesis:

Beetles and larvae prefer a light-colored substrate.

Methods:

Cover half of the floor of each petri dish with black construction paper and half with white construction paper. Tape the edges down so the beetles and larvae can't crawl under. Put one beetle in the middle of each dish. Wait a half hour or longer and record their choices. Then repeat the procedure with one larva per dish.

Result:

Your result is a statement of your beetles' choices and your larvae's choices. You can record your results as a histogram (see Figure 7.4 for an example of a completed histogram).

Conclusion:

The prediction is not confirmed. Adult beetles tend to prefer a dark substrate. This preference helps protect them from predators. My captive mantises have trouble spotting a beetle on a dark floor or the dark edge of a terrarium. When I put beetles in a terrarium with a dark edge, they run to the dark part and stay there. A beetle on the floor of an all-white container with a mantis is dead meat. In my experience the larvae have no preference between a light and dark substrate, which makes sense because they are equally visible on either.

Experiment 6

Question:

Are *Tenebrio* adults attracted to cold or warmth?

Hypothesis:

The beetles are attracted to warmth.

Methods:

Get a long baking tray and put a hot water bottle or a heating pad under one end of it, a tray of ice cubes under the other end. Wait a half-hour or more for the tray to get cool on one end and warm on the other. Put several beetles in the middle. Wait at least a half-hour or so to record the results.

Result:

Your result is a statement of your beetles' choices. You can record your results as a histogram (see Figure 7.4 for an example of a completed histogram).

Conclusion:

The prediction is not confirmed. Most beetles prefer the cool end.

Experiment 7

Question:

Do *Tenebrio* adults in a petri dish "bunch" (huddle together) or stay separate? Does moisture affect the outcome?

Hypothesis:

Beetles will not huddle together or bunch in a dry or moist container.

Methods:

Get two containers like petri dishes. Put a damp substrate in one and leave the other dry. If you use paper for the damp substrate, tape the edges down before adding moisture. Put two or more beetles into each container, without bran (see Figure 18.8). Wait a half hour or so to check on them. How many of them are in contact with another individual?

Result:

Your result is a statement of your animals' reaction in each dish.

Fig. 18.8 Jennifer recording observations of her beetles on a damp substrate.

Conclusion:

Adult beetles will usually bunch or huddle together in a dry dish, but not in a damp dish. Contact with another individual reduces water loss from body surfaces, by reducing the amount of body surface area exposed to the drying effects of air.

Experiment 8

Question:

Do *Tenebrio* adults like other foods as well as bran?

Hypothesis:

They might like cookies.

Methods:

Get a square box about 1 foot (30.4 cm) in diameter. Put one food item in each corner, for example, oat bran; wheat bran; sawdust; a crumbled cookie; flour; sugar; salt; loose, dry soil; or other things the students suggest. Give the beetles enough time so that most of them have gravitated to one pile or another.

Result:

Your result is a statement of your beetles' choices.

Conclusion:

Of the items suggested above, sugar and salt are the least acceptable to the beetles, in my experience. They will usually burrow into a pile of bran and sometimes into the others.

Experiment 9

Question:

Does a chilled beetle move more slowly than a warm beetle?

Hypothesis:

A chilled beetle will move more slowly than a warm beetle.

Methods:

Make a grid on a sheet of paper with squares about $1^1/_2$ inches (3.8 cm) in diameter. Put the paper inside a box. Put on the grid an adult beetle straight from your culture container. How many grid lines does the beetle cross in thirty seconds? Now replace that beetle with one that's been chilled in a freezer for about one minute (longer if it's chilled in a refrigerator) and repeat the procedure. (The time required to chill the beetle adequately may vary depending on the size of the container and the temperature of the freezer. Leaving it too long will kill it.)

Result:

Your result is a statement of the number of grid lines crossed by each beetle in thirty seconds.

Conclusion:

The prediction is confirmed. A chilled beetle crosses fewer grid lines than a warm beetle. Chilling reduces the body temperature of a cold-blooded (poikilothermic) animal and makes the animal move more slowly.

Experiment 10

Question:

Can a beetle walk faster on a slick surface or a rough surface?

Hypothesis:

A beetle can walk faster on a rough surface.

Methods:

Mark a starting point on two surfaces, one a very slick surface like smooth plastic and one a more textured surface like wood or smooth cloth. Put a beetle at each starting point, set a time, and poke both beetles from behind to get them moving. After thirty seconds or less, stop them and measure the distance covered by each beetle. Do beetles have trouble walking on slick surfaces?

Result:

Your result is a statement of the distance covered by each beetle.

Conclusion:

The prediction is confirmed. *Tenebrio* beetles have more trouble than many other insects walking on slick surfaces. They sometimes walk and walk, while not moving forward at all. They're adapted to living in grain and under bark, where slick surfaces are not an issue.

Experiment 11

Question:

Which is better at staying on its feet, a roly-poly or a *Tenebrio* beetle?

Hypothesis:

Roly-polies are better at staying on their feet.

Methods:

Put a beetle and a roly-poly in a flat, smooth-floored box. Which one turns over first? Begin again nine more times. Which one flips over first more often?

Result:

Your result is a statement of which one flips over more often and any relevant observations.

Conclusion:

The prediction is confirmed. *Tenebrio* beetles are the clumsiest animals I know of. They flip over constantly with no provocation and lie helplessly flailing the air with their legs. But in grain piles where they flourish, flipping over is not a problem. Roly-polies, on the other hand, are more accustomed to flat surfaces, at least on occasion, and so have evolved the ability to avoid constant flipping.

Postscript
Making Connections Among the Chapters

Lots of the creatures in this book share certain behaviors. That's to be expected, especially among those that share a habitat or share a food source. Most predators have certain things in common, as do most of the under-log creatures. Most of the latter are scavengers, eating decomposing plant or animal material.

Seeing the trends or common elements among chapters helps students extrapolate to creatures not in this book. If wolves and cheetahs both asphyxiate their prey before eating them, then maybe lions do too. Or if turtles and lizards both like to sun themselves, then maybe snakes do too.

Creatures Found in and Under Logs

The most outstanding common thread I see is among the under-log creatures. Coming from the same habitat, it's not surprising that they all prefer conditions like those under a log when offered a choice. I think what is surprising to children, or really to students of any age, is that these lowly creatures have "sense" enough to have any preference at all. I've never heard a child verbalize this, but I think most of us assume that creatures like insects are found where they are because that's where they hatched. Before I got to graduate school I'm sure I thought that. And it's true that most invertebrates don't really have the brain capacity to evaluate their options. But they possess an innate attraction to the conditions most suited to their own particular physiology. This prevents them from wandering away from the log where they hatched. Or it helps them find another log if they somehow get displaced.

What are the conditions under a log? Damp, dark, cool, covered. In the experiments in this book, four under-log creatures were offered a choice between a damp place and a dry place. Roly-polies, earthworms, crickets, and millipedes all prefer dampness. The millipede and earthworm chapters suggest offering a choice between cool and warm—both animals tend to prefer cool. Can the students predict what crickets and roly-polies might prefer? Three of the under-log animals show a preference for darkness. In general, the under-log creatures tend to choose the same conditions in which we find them in nature. See Figure 1, a table that summarizes the similarity of their choices. To fill out every block in the chart, you'll need to apply instructions from one chapter to another. For example, the roly-poly chapter does not have an experiment to offer them a choice between cool and warm, but the millipede chapter does, so apply the instructions for millipedes to roly-polies. Figure 2 is a blank table for the students to fill out. What would the students guess about under-log animals that aren't in the book, such as centipedes or snails or termites?

Flying Insects and Light

Some animals are attracted to light. Most of them are flying animals, such as fruit flies, moths, and flying beetles. One benefit of this attraction is that it causes them to walk out into the open or to climb where they may become airborne more easily. Can the students make predictions about other flying insects, such as houseflies, dragonflies, or adult ant lions?

Similarities and Differences Among Predators

There is a common element of prey recognition shared by most of the predators in this book. Most are generalist predators, which means they'll eat just about any prey that is small enough and not toxic. The praying mantises and the spiders are generalist predators, and they recognize an item as prey by the fact that it's moving. None of them will notice or attack an insect that isn't moving

Figure 1. Summary of choices made by under–log creatures. Choices may vary with species.

	Damp or Dry Substrate	Well-Lit or Deep Shadow	In House or Out of House	Cool or Warm	Light or Dark Substrate	Other
Roly-Polies	Damp	Dark	Not tested	Not tested	Not tested	
Earthworms	Damp	Dark	Not tested	Cool	Not tested	
Crickets	Damp	Prefer dark tube, but sometimes prefer well-lit end of terrarium	In	Not tested	Not tested	
Millipedes	Damp	Dark	In	Cool	Not tested	

Figure 2. For summarizing choices made by under–log creatures. Your results may vary considerably from mine, if you use different species of animals or if your conditions are slightly different.

	Damp or Dry Substrate	Well-Lit or Deep Shadow	In House or Out of House	Cool or Warm	Light or Dark Substrate	Other
Roly-Polies						
Earthworms						
Crickets						
Millipedes						

Which animals made choices that you might expect, given the conditions in their natural habitat?

_____

Which ones didn't? _____

Do you have any ideas about why they didn't? _____

regardless of how fresh it is. Ant lions too will not recognize prey that isn't moving, but will toss out a dead thing like it's a tiny pebble.

The ladybug beetle is the exception, and that's because ladybugs are not really generalist predators. Ladybugs eat soft-bodied insects, usually aphids and scale insects, neither of which move much. A ladybug will plow along a stem of aphids like a lawn mower, eating every one it comes to. The aphids just sit there, making no attempt to get out of the way. The ladybug recognizes its prey by some means other than movement.

Since they eat mostly aphids, they may recognize the aphids' shape, soft body, or taste. Ladybugs I have kept in captivity have refused fruit flies, which are about the same size as aphids.

Predators as specialized as ladybugs are unusual, because in nature it doesn't generally pay to be finicky. An eat-what's-available strategy usually pays off better for predators. Aphids are a relatively safe bet, though, because they are so abundant on their host plants, and the ladybug eggs are laid and hatch right on the same plants.

Can the students extrapolate to other predators that aren't in the book? What about predatory lizards and salamanders or predatory ground beetles (all generalists)? Do the students think they recognize prey by its movement?

Predatory behavior varies among animals in lots of other ways too. Some predators, such as ladybug beetles, wander around looking for prey. Others, like orb-weaving spiders, are sit-and-wait or ambush predators.

Some predators swallow prey whole, although none are in this book. But examples include frogs, toads, and lizards. Some, such as mantises, eat it in small pieces, while others, like ant lions and spiders, suck the insides out.

Some predators poison their prey, some don't. Some predators store prey.

Following is a list of questions and answers summarizing these similarities and differences (see Figure 3). Figure 4 is a set of the same questions without answers to photocopy for the students.

The Effects of Cold on Growth Rates

Chapters 8, 13, 14, and 16 all had experiments addressing the effect of either food or temperature on the growth rate of young. Since all the animals in this book are cold-blooded (all animals except birds and mammals are), all responded to cooler temperatures and reduced quantity of food by growing more slowly. Warm-blooded animals would be much more likely to respond to either with impaired health. Use Figure 5 to summarize your results.

Insect Reproduction

The animals in the last section of the book will all reproduce easily in captivity—fruit flies, aphids, and *Tenebrio* beetles. But their life cycles vary considerably. The flies and beetles lay eggs, but the aphids give birth to live young. The aphids, along with some other animals in the book, have gradual metamorphosis. Their young are nymphs, or miniature replicas of their parents, instead of larvae. The beetles and flies, and others earlier in the book, go through simple metamorphosis. The larva metamorphoses first into a pupa and then into an adult that looks completely different from the larva.

Which animals in the last section have the shortest life cycle? Which have the longest? Following this section you'll find a list of questions and answers to help the students compare these animals (see Figure 6). Figure 7 is a set of the same questions without answers to photocopy for the students.

Your students may see other trends. Encourage them to make other connections among chapters.

Photocopy Figure 4 and have students answer the following questions, choosing their answers from among these predators: wandering spiders, web-building spiders, ant lions, mantises, and ladybug beetles.

Questions/Answers

Photocopy Figure 4 and have students answer the following questions, choosing their answers from among these predators: wandering spiders, web-building spiders, ant lions, mantises, ladybug beetles, giant water bugs, backswimmers, diving beetles, dragonfly nymphs, toads, and frogs.

1. Which of the predators are generalists (will take almost any prey of the right size)?

   All except ladybug beetles.

2. Which of the predators are specialists (will take only one or maybe two prey types, rejecting others of similar size)?

   Ladybug beetles.

3. What advantage might there be in being a generalist predator instead of a specialist predator?

   A generalist seldom passes up available prey, so generalizing makes them more likely to get a meal.

4. What disadvantage might there be in being a generalist?

   A generalist predator might attack prey it can't handle or can't digest and perhaps be injured. It also must be equipped physiologically and anatomically to handle a variety of prey. A jack-of-all-trades is often not as good at any one job as a specialist.

5. What advantage might there be to a specialist predator?

   Its body is specialized physiologically and anatomically for catching and handling only one prey type. So it probably utilizes the prey type more efficiently than a generalist would.

6. What disadvantage might there be to a specialist predator?

   If its prey species disappears, it's in trouble. That's why you find few specialist predators, and their prey is generally superabundant (or was, before habitat loss and climate change began to impact wildlife populations).

7. Did any predators continue to take one prey after another for an extended period, without refusing any?

   Ladybugs and ant lions will eat for a long time, because their prey are so small. Mantises will also eat small prey for a long time and will sometimes continue to strike even when full, eating only part of the prey, then dropping it. Spiders often wrap prey and save them for later when full.

8. Which predators ate only moving prey?

   Mantises, ant lions, most spiders.

9. Did any accept dead prey?

   Mantises, when prey is jiggled (sometimes).

10. Which predators did you see wander around looking for prey?

    Wandering spiders and ladybug beetles. Mantises sometimes pursue prey a short distance once they see it, but don't wander much.

11. Which predators seemed to be mainly sit-and-wait or ambush predators?

    Ant lions and web-building spiders; mantises, most of the time.

12. What advantage might wandering predators have over ambush predators?

    They encounter more prey.

13. What disadvantage might wandering predators have?

    They expend more energy than ambush predators and also make themselves much more visible to larger predators who might eat them.

14. What advantage might there be to the sit-and-wait strategy?

    Ambush predators expend less energy and keep themselves more concealed from larger predators than do the wandering predators.

15. What disadvantage might there be to the sit-and-wait or ambush strategy?

    Ambush or sit-and-wait predators encounter fewer prey than wandering predators.

16. Which prey have traps?

    Ant lions and web-building spiders.

17. Are those with traps sit-and-wait predators?

    Yes.

18. What disadvantages and advantages are there to traps?

    They take a lot of energy to construct and repair. They restrict the predator to prey in one small area. The advantage is that they are very efficient at catching whatever prey do come into that small area.

19. Which predators ate the whole body of the prey, but piece by piece?

    Mantises, ladybug beetles.

20. Which predators sucked out the prey's insides, leaving an empty carcass?

    Ant lions, spiders.

21. Which predators inject poison and/or digestive juices into their prey?

    The same ones that suck out their insides.

22. Do any predators store their prey for the future?

    Web-building spiders.

23. Which predators did you like best? Why?

24. Can you think of other similarities and differences among the predators?

Figure 3 Looking for similarities and differences among predators (Answer guide).

Name: _____   Date: _____

Answer the following questions, choosing your answers from among these predators:

- **wandering spiders**   • **web-building spiders**   • **ant lions**   • **mantises**
- **ladybug beetles**   • **giant water bugs**   • **backswimmers**   • **diving beetles**
- **dragonfly nymphs**   • **toads**   • **frogs**

1. Which of the predators are generalists (will take almost any prey of the right size)?

2. Which of the predators are specialists (will take only one or maybe two prey types, rejecting others of similar size)?

3. What advantage might there be in being a generalist predator instead of a specialist predator?

4. What disadvantage might there be in being a generalist?

5. What advantage might there be to a specialist predator?

6. What disadvantage might there be to a specialist predator?

7. Did any predators continue to take one prey after another for an extended period, without refusing any?

8. Which predators ate only moving prey?

9. Did any accept dead prey?

10. Which predators did you see wander around looking for prey?

11. Which predators seemed to be mainly sit-and-wait or ambush predators?

12. What advantage might wandering predators have over ambush predators?

Figure 4 Looking for similarities and differences among predators in the book.

13. What disadvantage might wandering predators have?

14. What advantage might there be to the sit-and-wait strategy?

15. What disadvantage might there be to the sit-and-wait or ambush strategy?

16. Which prey have traps?

17. Are those with traps sit-and-wait predators?

18. What disadvantages and advantages are there to traps?

19. Which predators ate the whole body of the prey, but piece by piece?

20. Which predators sucked out the prey's insides, leaving an empty carcass?

21. Which predators inject poison and/or digestive juices into their prey?

22. Do any predators store their prey for the future?

23. Which predators did you like best? Why?

24. Can you think of other similarities and differences among the predators?

Figure 4b Looking for similarities and differences among predators in the book.

Name: _____   Date: _____

In which two chapters did you find an effect of cold on the growth rate of the young?

       Chapter 15: Baby mantises _____

       Chapter 16: Fruit flies _____

       Chapter 18: Mealworms _____

       Other: _____

If you made graphs for each of the above experiments, which graph showed the greatest effect of cooler temperatures?

Which insect's growth rate (if any) showed the greatest effect of varied food availability?

Why would a baby mammal or bird placed in a cooler environment without protection (no mom or thick fur) be more likely to be harmed by the cooler temperatures? (NOTE: Don't try this!)

Figure 5 Comparing results of the experiments that tested the effects of cooler air or varied food availability on growth rates.

Photocopy Figure 7 and have students answer the following questions:

1. Which five of these seven animals lay eggs that hatch into larvae: fruit flies, aphids, *Tenebrio* beetles, ladybug beetles, black swallowtail butterflies, monarchs, hissing roaches?
    Fruit flies, all the beetles, both butterflies.

2. Which two do not lay eggs?
    Aphids, hissing roaches.

    What do they do instead?
    Aphids and hissing roaches give birth to live young.

3. Which two have young that look like miniature adults, instead of larvae?
    Aphids and hissing roaches.

    What are such young called?
    Nymphs.

4. Do larvae usually look like their parents?
    No.

5. Do nymphs usually look like their parents?
    Yes.

6. Do larvae pupate and metamorphose?
    Yes. Their development is called simple or complete metamorphosis.

7. Do nymphs pupate?
    No, they molt several times as they grow, but there is no abrupt major change in body form. Their development is called gradual metamorphosis.

8. Can you name another animal in this book that has nymphs instead of larvae?
    Crickets, mantises.

9. What other animals in this book have young called larvae?
    Ant lions, bessbugs.

10. Which animals in Chapters 16, 17, and 18 had the fastest development time (egg or birth to adult)?
    Aphids.

11. Which had the longest development time?
    Tenebrio beetles.

12. What advantage is there to having larvae (called simple or complete metamorphosis)?
    Because the larvae are different physically, they often have a different diet and habitat than the adult, one where food may be more plentiful so the young can grow fast. So in a way, the larvae are specialized for exploiting a habitat where food is plentiful for growth, while adults are specialized for dispersal or traveling to new areas.

    Animals with gradual metamorphosis must accomplish both tasks with one body form. The advantage of gradual metamorphosis is that they avoid pupation, which takes a lot of energy and time and often leaves the animal vulnerable to predators. Pupae are usually immobile.

Figure 6 Insect Reproduction

Name: _____   Date: _____

1. Which five of these seven animals lay eggs that hatch into larvae?
   - fruit flies
   - aphids
   - *Tenebrio* beetles
   - ladybug beetles
   - black swallowtail butterflies
   - monarchs
   - hissing roaches

2. Which two do not lay eggs?

   What do they do instead?

3. Which two have young that look like miniature adults, instead of larvae?

   What are such young called?

4. Do larvae usually look like their parents?

5. Do nymphs usually look like their parents?

6. Do larvae pupate and metamorphose?

7. Do nymphs pupate?

8. Can you name another animal in this book that has nymphs instead of larvae?

9. What other animals in this book have young called larvae?

10. Which animals in Chapters 16, 15, and 18 had the fastest development time (egg or birth to adult)?

11. Which had the longest development time?

12. What advantage is there to having larvae (called simple or complete metamorphosis)?

Figure 7 Insect Reproduction

Appendix

Information

For information on the conservation of invertebrates and other wildlife, check out the website of teh Center for Biological Diversity (www.biologicaldiversity.org). Click on "invertebrates" under the "species" tab.

For information on how you can get involved in the push for renewable energy and the fight against global warming (both crucial issues to the survival of wildlife), I recommend several organizations that I've been involved in. There are, of course, many others!

- Greenpeace, at www.greenpeace.org/international/en/campaigns/climate-change/

- 350, at 350.org

- bugguide.net is helpful for bug identification

In addition, you can learn about the rampant illegal trafficking in wildlife and wildlife parts, and what you can do to stop it at "Traffic: The Wildlife Trade Monitoring Network," at www.traffic.org.

Ordering

Below are the names of four biological-supply and scientific-equipment companies that are widely known among science teachers. I have used Carolina Biological only because they are nearby. Of course you can find other sources by searching yourself, using terms such as "insect sleeve cage" or "petri dishes," etc.

- Carolina Biological Supply Company (North Carolina)

- Ward's Science (New York and California)

- Frey Science (New Hampshire)

- Nasco Science (Wisconsin and California)

Ladybug beetles are available from www.gardensalive.com, as well as many other gardening-supply websites (especially those that focus on organic or pesticide-free gardening).

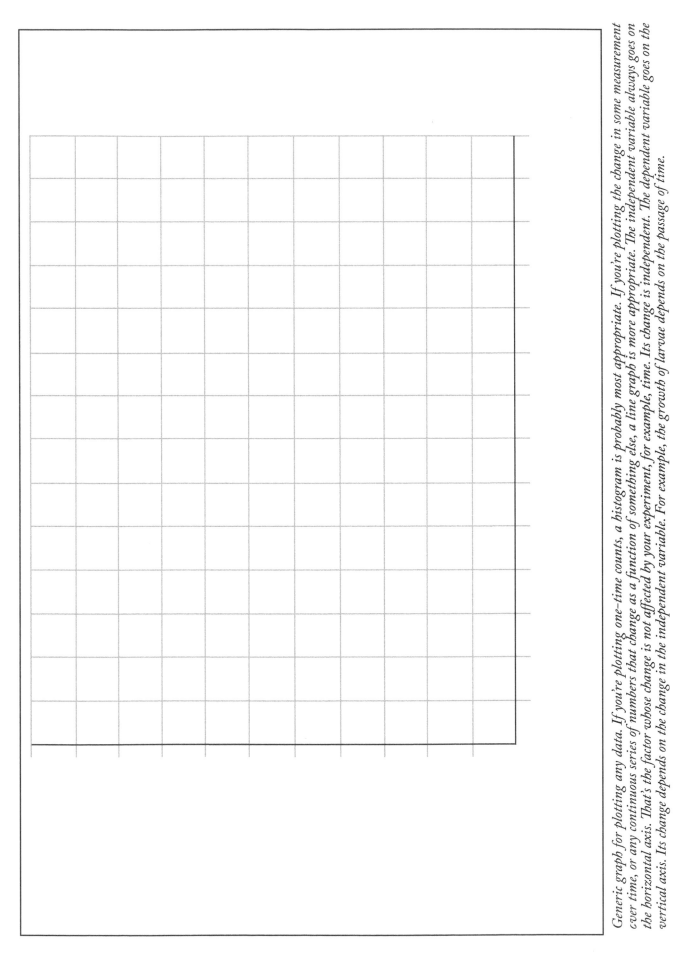

Generic graph for plotting any data. If you're plotting the change in some measurement over time, or any continuous series of numbers that change as a function of something else, a line graph is more appropriate. The independent variable always goes on the horizontal axis. That's the factor whose change is not affected by your experiment, for example, time. Its change is independent. The dependent variable goes on the vertical axis. Its change depends on the change in the independent variable. For example, the growth of larvae depends on the passage of time.

A generic table for recording any results.

Bibliography

The bibliographical listings for the chapters below are from the first edition, except for the new chapters. For the new chapters (Chapters 3, 4, 5, and 6), my sources were my own experience and the experience of friends. I did create a list of resources for each of those chapters, for the benefit of readers, which can be found at the end of each of those chapters.

Chapter 2—Attracting and Maintaining Critters

Barnes, Robert D. 1974. *Invertebrate Zoology*. W. B. Saunders Co., Philadelphia, London, Toronto.

Borror, Donald J., and Richard E. White. 1970. *A Field Guide to the Insects of America North of Mexico* (Peterson Field Guide series). Houghton Mifflin Co., Boston.

Carolina Biological Supply Company. 1990–91. *Biology/Science Materials: Catalog 61*. Carolina Biological Supply Co., Burlington, NC.

Conant, Roger. 1975. *A Field Guide to Reptiles and Amphibians of Eastern and Central North America* (Peterson Field Guide series). Houghton Mifflin Co., Boston.

Lawrence, Gale. 1984. *A Field Guide to the Familiar*. Prentice-Hall, Inc., Englewood Cliffs, NJ.

Roth, Charles E. 1982. *The Wildlife Observer's Guidebook*. Prentice-Hall, Inc., Englewood Cliffs, NJ.

Stokes, Donald W. 1983. *A Guide to Observing Insect Lives*. Little, Brown and Co., Boston, Toronto.

White, Richard E. 1983. *A Field Guide to the Beetles of North America* (Peterson Field Guide series). Houghton Mifflin Co., Boston.

Chapter 3—Beautiful and Big: Black Swallowtail Butterflies

See Additional Resources for Black Swallowtails at the end of Chapter 3.

Chapter 4—Mystical, Magical Monarchs Need Our Help

See the websites described at length in the chapter for additional resources.

Chapter 5—Bessbugs: Glossy and Glamorous

See Additional Resources for Bessbugs at the end of Chapter 5.

Chapter 6—Madagascar Hissing Roaches: Simply Stunning

See Additional Resources for Madagascar Hissing Roaches at the end of Chapter 6.

Chapter 7—Roly-Poly or Pill Bug Science

Levi, H. W., and L. R. Levi. 1968. *Spiders and Their Kin* (Golden Field Guide series). Golden Press, Western Publishing Co., Racine, WI. This book has two pages in the back on pill bugs, sow bugs, and the other "wood lice."

Raham, Gary. 1986. Pill bug biology: A spider's spinach, but a biologist's delight. *The American Biology Teacher* 48(1): 9–16.

Chapter 8—Wiggly Earthworms

Andrews, William A. 1973. *A Guide to the Study of Soil Ecology*. Prentice-Hall, Inc., Englewood Cliffs, NJ.

McLaughlin, Molly. 1986. *Earthworms, Dirt, and Rotten Leaves: An Exploration in Ecology*. Atheneum, New York.

Chapter 9—Chirping Crickets

Alexander, Richard D. 1961. Aggressiveness, territoriality, and sexual behavior in field crickets (Orthoptera: Gryllidae). *Behavior* 17: 130–220.

Andrews, William A. 1974. *A Guide to the Study of Terrestrial Ecology*. Prentice-Hall, Inc., Englewood Cliffs, NJ.

Borror, Donald J., and Richard E. White. 1970. *A Field Guide to the Insects of America North of Mexico* (Peterson Field Guide series). Houghton Mifflin Company, Boston.

Carolina Biological Supply Company. 1982. *Carolina Arthropods Manual*. Carolina Biological Supply Co., Burlington, NC.

Milne, Lorus, and Margery Milne. 1980. *The Audubon Society Field Guide to North American Insects and Spiders*. Alfred A. Knopf, Inc., New York.

Wyler, Rose. 1990. *Grass and Grasshoppers*. Messner, Englewood Cliffs, NJ.

Zim, Herbert S., and Clarence Cottam. 1987. *Insects: A Guide to Familiar American Insects*. Golden Press, New York.

Chapter 10—Many-Legged Millipedes

Barnes, Robert D. 1974. *Invertebrate Zoology*. W. B. Saunders Co., Philadelphia, London, Toronto.

Levi, H. W., and L. R. Levi. 1968. *Spiders and Their Kin* (Golden Field Guide series). Golden Press, Western Publishing Co., Inc., Racine, WI.

Preston-Mafham, Ken. 1990. *Discovering Centipedes and Millipedes*. Bookwright Press, New York.

Chapter 11—The Slime That Creeps

Carolina Biological Supply Company. 1986. *Techniques for Studying Bacteria and Fungi*. Carolina Biological Supply Co., Burlington, NC.

Ellis, Harry. 1992. It's slimy! It's alive! *Wildlife in North Carolina* 56(6): 24–28.

Farr, M. L. 1981. *How to Know the True Slime Molds*. William C. Brown Co., Dubuque, IA.

Lincoff, Gary H. 1981. *The Audubon Society Field Guide to North American Mushrooms*. Alfred A. Knopf, Inc., New York. This has an excellent and thorough section on identifying slime molds, with color photographs of plasmodia and sporangia.

Stephenson, S. L. 1982. Slime molds in the laboratory. *American Biology Teacher* 44(2): 119–120, 127.

Stephenson, S. L. 1985. Slime molds in the laboratory II: Moist chamber cultures. *American Biology Teacher* 47(8): 487–489.

Stevens, R. B., ed. 1981. *Mycology Guidebook*. University of Washington Press, Seattle.

Chapter 12—Ant Lions: Terrors of the Sand

Banks, Joan Brix. 1978. Ant lion. *Ranger Rick's Nature Magazine* 12(8): 28–29.

Burton, Maurice, and Robert Burton. 1974. *Funk and Wagnalls Wildlife Encyclopedia*. Funk and Wagnalls, Inc., New York.

Zim, Herbert S., and Clarence Cottam. 1987. *Insects: A Guide to Familiar American Insects*. Golden Press, New York.

Chapter 13—Spooky Spiders

Comstock, J. H. 1948. *The Spider Book*. Comstock Publishing Co., Inc., Ithaca, NY.

Kaston, B. J. 1972. *How to Know the Spiders*. William C. Brown Co., Dubuque, IA.

Land, M. F. 1971. Orientation by jumping spiders in the absence of visual feedback. *Journal of Experimental Biology* 54: 119–139.

Lavine, S. A. 1966. *Wonders of the Spider World*. Dodd, Mead and Co., New York.

Levi, H. W., and L. R. Levi. 1968. *Spiders and Their Kin* (Golden Guide Field Guide series). Golden Press, Western Publishing, Inc., Racine, WI.

Milne, Lorus, and Margery Milne. 1980. *The Audubon Society Field Guide to North American Insects and Spiders*. Alfred A. Knopf, Inc., New York.

Moffett, Mark W. 1991. All eyes on jumping spiders. *National Geographic*, September.

Rovner, J. S. 1989. Wolf spiders lack mirror-image responsiveness seen in jumping spiders. *Animal Behavior* 38: 526–533.

Chapter 14—Praying Mantises: The "Smartest" Insects

Lavies, Bianca. 1990. *Backyard Hunter: The Praying Mantis*. Dutton, New York. This book has fantastic photographs of every life stage.

Milne, Lorus, and Margery Milne. 1980. *The Audubon Society Field Guide to North American Insects and Spiders*. Alfred A. Knopf, Inc., New York. This book can be useful in distinguishing species.

Zim, Herbert S., and Clarence Cottam. 1987. *Insects: A Guide to Familiar American Insects*. Golden Press, New York.

Chapter 15—Baby Mantises (Nymphs)

Conklin, Gladys. 1978. *Praying Mantis, the Garden Dinosaur*. Holiday House, New York.

Johnson, Sylvia A. 1984. *Mantises*. Lerner Publications. Minneapolis.

Lavies, Bianca. 1990. *Backyard Hunter: The Praying Mantis*. Dutton, New York.

Chapter 16—The Two-Week Life Cycle of Fruit Flies

Carolina Biological Supply Company. 1982. *Carolina Arthropods Manual*. Carolina Biological Supply Co.,

Burlington, NC. This booklet tells how to culture houseflies indoors (get them to reproduce indefinitely) under sanitary conditions.

Carolina Biological Supply Company. 1988. *Carolina* Drosophila *Manual*. Carolina Biological Supply Co., Burlington, NC.

Cunningham, John D. 1970. *First You Catch a Fly*. McCall Publishing Co., New York. This book is full of ideas about things to do with fruit flies and houseflies.

Milne, Lorus, and Margery Milne. 1980. *The Audubon Society Field Guide to North American Insects and Spiders*. Alfred A. Knopf, Inc., New York. Look up *Drosophila*, not fruit fly.

Chapter 17—Aphids and Their Predators, Ladybug Beetles

Burton, Maurice, and Robert Burton. 1974. *Funk and Wagnalls Wildlife Encyclopedia*. Funk and Wagnalls, Inc., New York.

Johnson, Sylvia A. 1983. *Ladybugs*. Lerner Publications, Minneapolis.

Watts, Barrie. 1987. *Ladybug*. Silver Burdett, Morristown, NJ.

White, Richard E. 1983. *A Field Guide to the Beetles of North America* (Peterson Field Guide series). Houghton Mifflin Co., Boston.

Zim, Herbert S., and Clarence Cottam. 1987. *Insects: A Guide to Familiar American Insects*. Golden Press, New York.

Chapter 18—Insect Metamorphosis on a Macro Scale: Mealworms to Beetles

Carolina Biological Supply Company. 1982. *Carolina Arthropods Manual*. Carolina Biological Supply Co., Burlington, NC.

White, Richard E. 1983. *A Field Guide to the Beetles of North America* (Peterson Field Guide series). Houghton Mifflin Co., Boston.

Index